Chronic Disease
in the Twentieth Century

Chronic Disease in the Twentieth Century

A HISTORY

George Weisz

Johns Hopkins University Press
Baltimore

Johns Hopkins University Press

2715 North Charles Street

Baltimore, Maryland 21218-4363

www.press.jhu.edu

Library of Congress Cataloging-in-Publication Data

Weisz, George, author.

   Chronic disease in the twentieth century / George Weisz.

     p. ; cm.

   Includes bibliographical references and index.

   ISBN 978-1-4214-1302-0 (hardcover : alk. paper) — ISBN 1-4214-1302-7
(hardcover : alk. paper) — ISBN 978-1-4214-1303-7 (pbk. : alk. paper) —
ISBN 1-4214-1303-5 (pbk. : alk. paper) — ISBN 978-1-4214-1304-4
(electronic) — ISBN 1-4214-1304-3 (electronic)

   I. Title.

   [DNLM: 1. Chronic Disease. 2. Health Policy—history. 3. History, 20th
Century. 4. Public Health Practice—history. WT 11.1]

   RA418

   362.1—dc23   2013027737

A catalog record for this book is available from the British Library.

*Special discounts are available for bulk purchases of this book. For more
information, please contact Special Sales at 410-516-6936 or specialsales@press
.jhu.edu.*

Contents

Preface

The structure of this book reflects my conviction that chronic disease—though a traditional medical term—was primarily an American *policy* construct during the first half of the twentieth century. In part I, I devote seven chapters to the United States and offer what I hope is a somewhat different perspective on the history of American healthcare from that offered by work focused on political battles over health insurance. I argue that the concept of "chronic disease" was developed in three domains during the interwar period. By the mid-1950s, two transformative developments had made the concept central to the rhetoric and organization of public health, healthcare, and research, and allowed individuals in different domains, for by now there were more than three, to come together and negotiate the meanings of and actions around chronic disease. In part II, three chapters discuss the very different British and French approaches to chronic disease.

Chapter 1 deals with the first of these domains, which appeared around the time of World War I. Its earliest manifestation was concern about chronic illness that emerged within New York City's Health Department. This concern was justified by apparently rising mortality rates for illnesses like cancer, an increase that was in fact hotly contested by many statisticians, who viewed it as an artifact of changing disease categories and diagnostic procedures. (This debate, which continued to be a divisive issue, recurs throughout this book.) But the wider spread of a chronic disease problem owed much to the movement for periodic physical examinations that emerged from the insurance industry and eugenic movement. In 1908 the economist Irving Fisher wrote a report on "National Vitality" for President Theodore Roosevelt's Conservation Commission, in which he argued that there was huge waste of American life and health as a result of failure to make use of existing knowledge about diseases. Several years later Fisher was involved in the creation of the Life Extension Institute (LEI), a business enterprise that performed medical examinations for many clients, but notably life insurance companies with an interest in keeping their clients alive as long as possible. The American Medical Association (AMA) and the public health movement quickly adopted the principle of the periodic physical examination, even if the practice failed to achieve mass popularity. Although all incipient diseases and "defects" might be detected and remedied, degenerative chronic diseases were especially appropriate targets for examination because they were frequently "silent" and without symptoms until they became too advanced to treat.

Chapter 2 discusses a second domain in which the public health movement sought to expand its reach beyond sanitation and the prevention of infectious diseases to the prevention *and care* of *all* illnesses. Small, disorganized, and focused on a narrow range of problems, the public health movement began to evolve in the 1920s. By the middle of that decade, influential voices, notably that of Charles Winslow of Yale, began calling for expansion beyond traditional public health activities to include care of degenerative diseases like cancer, diabetes, and heart disease. New measures, he argued, were needed to face new conditions. This was a controversial position because it entailed a major expansion of the scope of public health. It also involved breaking down the distinction between prevention and cure; this was already occurring in public programs dealing with venereal disease and tuberculosis, which sought to find and cure individuals with these diseases but which could be justified as public health measures to prevent the spread of communicable diseases. Creating similar programs solely to prevent disease from reaching the stage of incurability was a new and contentious idea. By the mid-1930s, however, prestigious voices within public health institutions were making such arguments and referring to the new category "chronic disease," which was to be at the center of the expanded public health movement.

Thus, early concern about chronic disease associated with periodic medical examinations combined with a push by public health professionals to extend their reach. In chapter 3, I discuss a third domain and the one most directly devoted to chronic care, which emerged during the 1920s in the world of social welfare institutions, including hospitals, dealing with the destitute poor. Doctors working with this population like Ernst Boas, then medical director of the Montefiore Hospital for Chronic Diseases in the Bronx, hospital administrators in public hospitals seeking to free beds for acutely ill patients, and social welfare workers trying to place sick and infirm people in appropriate institutions dominated this movement in its early years. A series of local health surveys of welfare cases in various cities and states showed that a very large proportion of those dependent on welfare institutions suffered from chronic diseases or disabilities and that almshouses where they traditionally resided were inadequate and frequently inhumane. The goal of this reform movement was to create an appropriate and rational system of care for "indigent chronics" that included visiting nursing care in the home, purely custodial care, skilled nursing care, and sophisticated medical care when necessary. The most striking example of this welfare form of chronic care occurred in New York City, where an extraordinary array of reports, services, and institutions was organized, culminating in the opening in 1940 of the huge Goldwater Chronic Disease Hospital on Welfare (now Roosevelt) Island.

Chronic illness expanded into the national arena during the mid-1930s. In chapter 4 I analyze one transformational moment in this shift. New Deal reformers—including many New Yorkers who brought their city's welfare concerns to Washington—needed to demonstrate that the Depression had created terrible health conditions in order to advocate effectively for a broad transformation of American healthcare. They could not utilize mortality statistics, however, which were improving. A group led by pioneering epidemiologist Edgar Sydenstricker stumbled on the concept of "days of disability" developed in unemployment surveys and utilized it as a measure of illness status. No one had more days of disability than people who were chronically ill and disabled, and the result was the massive National Health Survey of Chronic Illness and Disability (NHS) undertaken in 1935–36. This survey purported to show that more than one in six Americans suffered from a chronic disease or disability. These figures were highly dubious and, although they were eventually revised, the original dramatic figures somehow stuck. The immediate effect of the survey was to transform chronic disease into a national social problem.

During the two decades following World War II, chronic illness emerged as a center of American health policy. In chapter 5, I examine a second transformative development, an unprecedented alliance of public and private institutions that focused on chronic illness, largely as a way of creating consensus around healthcare reform that avoided the virulent disagreements surrounding national health insurance. This alliance was embodied by the Commission on Chronic Illness (CCI, 1947–56), which brought together some of the major institutions of American healthcare. The commission publicized the problem in entirely new ways and published a major report that shaped American health policy into the 1980s. Avoiding issues that provoked serious conflict, notably national health insurance and health finances, the commission's report focused on what were by now areas of consensus: the need for massive investment in research on chronic disease, periodic physical examinations, and facilities for rehabilitation to get people back in the workforce or off welfare. But there were also major innovations: insisting on prevention rather than cure; keeping the chronically ill at home through subsidies and homecare instead of institutionalizing them; providing necessary medical care in new chronic-care wards in general hospitals instead of in remote and isolated chronic-care hospitals; building nursing homes for those unable to live at home but not requiring intense medical care; developing mass multiphasic screening technologies to detect diseases cheaply in their early and treatable stages. The commission's program was extremely ambitious and in many ways imprecise. But it exerted immense influence on the development of healthcare policy during the 1960s and 1970s.

In chapters 6 and 7, the discussion moves from theory to practice. Much of the history of American healthcare in the 1960s and 1970s reflected the recommendations of the Commission on Chronic Illness, in a context where federal funding was predominantly directed at groups that suffered disproportionately from chronic disease and disability—poor, elderly, and disabled people. Not surprisingly, efforts to implement the commission's recommendations did not go smoothly. I focus on the two areas that the commission itself emphasized. Chapter 6 looks at the development of medical and care facilities, from chronic-care wards in general hospitals to homecare programs. Chapter 7 examines how public health institutions in the United States responded to the chronic disease issue. My goal is not to judge whether such efforts were successful or not, but rather to analyze how the working out of chronic disease policy during the 1960s and 1970s reflected efforts to adapt the principles of the Commission on Chronic Illness to increasingly complex and intractable social and economic realities. In the case of care facilities, these included the aging of the population; the unexpected costs of all solutions, including homecare; the limitations of rehabilitation; and the difficulties in developing more than a handful of effective screening mechanisms. In the case of public health, problems included the complications of acting in a highly decentralized environment, limited funding, and the tension between disease-based and comprehensive programs.

The three chapters in part II focus on the United Kingdom and France and have two aims: (1) to pinpoint the conditions in the United States that made the chronic disease movement possible and whose absence in Europe led authorities there to confront problems differently; and (2) to analyze the health system rationales that framed problems in ways that had little to do with "chronic disease."

In both countries the notion of "chronic disease" eventually did spread, starting with the World Health Organization's (WHO) European Section's first meetings on the subject in the mid-1950s. Yet it only began to truly influence national policy in these countries several decades later.

Chapter 8 briefly describes key differences between the two European nations and the United States during the interwar and postwar periods. It is clear that appreciably older populations in the United Kingdom and France made chronic disease there appear to be essentially a problem of elderly people. This perception made it difficult to justify major investments in chronic disease programs as a way to increase national productivity and power. Economic devastation after World War II considerably narrowed choices that were available to health planners. Finally, the decision to create a National Health Service (NHS) in the United Kingdom and a national health insurance system in France required massive resources and forced authorities to take account of broad population needs rather than focus on a category like chronicity that primarily affected specific groups.

Chapter 9 examines the British case. During the interwar years, there were clear parallels with the United States: a belief in the potential of medical science to prevent and cure disease; support for research institutions; robust epidemiological work; and the development of what came to be called rehabilitation. Care of the chronically ill was discussed sporadically but never as a serious issue. Allegedly rising mortality rates from diseases like cancer were usually explained away. The most critical difference had to do with context. American reformers functioned in a geographically and institutionally fragmented environment requiring broad alliances to achieve institutional change; they characteristically did so by emphasizing how terrible the health situation was and by focusing on groups with longstanding entitlements to care: children especially, but also the welfare poor and, from 1935, older persons. Health reform in Britain, in contrast, was well underway during the interwar period.

National health insurance for working men, introduced before World War I, extended state-funded healthcare beyond the very poor. The Local Government Act of 1929 dismantled the Poor Law and allowed local authorities to take over Poor Law hospitals that contained a majority of the nation's hospital beds. This law theoretically ended the distinction between indigent and non-indigent hospital patients. Agents of public health, the medical officers of health were increasingly occupied with administering these enlarged local health institutions, thus erasing the distinction between prevention and cure. The Radium Trust was centrally organizing the distribution of radium for therapeutic purposes and was, in the process, setting standards for cancer therapy. There was little pressure to promote reform by emphasizing the inadequacies of chronic care. On the contrary, the tendency was to emphasize successes. Furthermore, national renewal was associated almost exclusively with the young. No one took much notice of infirm or chronically ill elderly people, who tended to remain in the dilapidated poorhouses that local authorities refused to take over. The British situation changed when the Ministry of Health became more actively involved in hospital planning during and after World War II. Surveys began to reveal the inadequacies of care for older ill and disabled individuals. While these people were still thought of as "bed-blockers," they developed into a social and humanitarian problem. The most conspicuous result of this movement was the development of geriatrics as a hospital specialty. By the 1960s, something like the American understanding of chronic disease appeared through the work of epidemiologists like Jerry Morris. But it did not become a central category of health policy during most of the twentieth century.

Chapter 10 outlines the French situation, which contrasts even more sharply with that of the United States. As in the United Kingdom, chronicity was associated with the indigent elderly population. Until the 1960s it remained a secondary administrative matter since the focus of population policy had for almost a century been raising the birth rate and improving maternal-infant care. With an expanding health insurance system to finance, no one was much interested in discovering new problems. Insurance reimbursement of long-term care was generous, but in a hospital-oriented environment, chronic illness was largely a nuisance. There existed a system of hospices to care for those who could not be helped by medical care, but the reality was that chronic patients always took up hospital beds. Parisian administrators dealt with the problem by creating special chronic-care hospitals in remote suburbs; these had minimal medical facilities, were much cheaper to operate, and freed up beds in central Paris. A handful of individuals in France sought to make chronic disease more central to healthcare. But by the 1970s, the issue of chronic disease had been displaced by two developments. First, old age erupted in the public arena as a social problem that was only partly medical. The central difficulty of elderly people was seen to be social "exclusion," a notion that lumped them together for policy purposes with those who were disabled, mentally ill, homeless, and unemployed. Second, reform of hospitals, which began in 1970 and has continued since, centered attention on a series of struggles over power and structure. Reform was also deliberately accompanied by a change of vocabulary. Rather than terms like *chronic* and *hospice* with their stigmatizing connotations, the new nomenclature focused on length of stay: hospitals and related institutions were defined as short-term, mid-term, and long-term and major disagreements had to do with whether certain long-term institutions were adequate and whether patients were in an inappropriate category of institution. These developments eclipsed talk about a chronic disease problem for much of the twentieth century.

In the epilogue, I briefly examine the current situation. By the twenty-first century, chronic disease had become a central policy issue in all three countries, in virtually all developed nations, and, to some degree, in the developing world as well. Much of it has to do with the fact that countries in the developed world face common problems that have expanded exponentially: a growing population of middle-aged and elderly persons who place new demands on public facilities and finances and who are arguably not well served by the current system of private medical practice dominated by specialists. But problems in different countries have always been broadly similar; what diverged were the approaches taken to deal with them. It is these approaches that have begun rapidly to converge. One reason is the perceived financial crisis facing virtually every healthcare system. Another is a developing international conversation in which many countries and international organizations participate. The World Health Organization has played a major role in globalizing the problem and suggesting solutions. The European Community provides another arena for discussion of and research in a transnational context. But if everyone now seems to acknowledge the existence of a chronic disease crisis and is searching for better forms of "chronic disease management" using broadly similar terms and concepts, the different paths taken by each country during the twentieth century continue to exert profound influence.

This book was conceived in Paris in 2005–6 during my yearlong joint Canadian Institutes of Health–INSERM fellowship as a visiting researcher. I am grateful to both organizations for making that extraordinary year possible. While in Paris, I was attached to CERMES 3, a research unit that I have for some time regarded as my second institutional home. I am grateful to CERMES 3 director Martine Bungener and her successor, Jean-Paul Gaudillière, for making that and succeeding stays in Paris so pleasant and productive. Among the scholars with whom I discussed my initial confusion and developing ideas about chronic disease were the sociologists Isabelle Baszanger and Nicolas Henckes, who continued to discuss and question my work as I proceeded on what turned out to be a long journey of discovery. They co-organized with me (and did the lion's share of the work on) an international workshop on chronic disease that took place in Paris in 2011. That meeting taught me much and helped me clarify my arguments. Isabelle and Nicolas, Virginia Berridge, Martin Gorsky, and Paul Bridgen all took time out from busy schedules to read and comment perceptively on individual chapters.

I was fortunate enough to receive funding for this project from the Social Science and Humanities Research Council of Canada, the Canadian Institutes of Health Research, and Le Fonds Québécois de la Recherche sur la Société et la Culture. I am grateful to all three institutions for their support, which enabled me to work with several gifted research assistants, including Jonah Campbell, Donna Evleth, Zev Moses, Tess Lanzarotta, and Kat Duckworth. The funding also permitted me to do research in numerous archives, including those at Yale University, the American Philosophical Society, the US National Archives, the National Library of Medicine (NLM), the University of Minnesota, Columbia University, the New York Academy of Medicine (NYAM), the Assistance Publique (Paris), and the Ministère de la Santé (Paris). I am grateful to archivists in all these institutions who have been unfailingly helpful. But special thanks must go to Stephen Greenberg of the NLM and Arlene Shaner of NYAM for their guidance and help. Dr. Carla Keirns of Stony Brook University provided me with documents that had been inaccessible to me.

Part of chapter 4 was previously published in an earlier form in 2011 as "Epidemiology and Health-Care Reform: The National Health Survey of 1935–1936," *American Journal of Public Health* 101 (3): 438–47 by the Sheridan Press.

Authors are not easy people to live with and I am more difficult than most. I am thankful to Zeeva for putting up with me and for nearly always finding the objects that I regularly mislay.

Abbreviations

ACS	American College of Surgeons
ADL	activities of daily living
AHA	American Hospital Association
ALDs	*affections de longue durée*
AMA	American Medical Association
AP	Assistance Publique
APHA	American Public Health Association
APWA	American Public Welfare Association
ASCC	American Society for the Control of Cancer
BMA	British Medical Association
CCI	Commission on Chronic Illness
CCM	chronic care model
CCMC	Committee on the Costs of Medical Care
CDC	Centers for Disease Control
CDSC	Communicable Disease Surveillance Centre
CHP	Department of Chronic Diseases and Health Promotion
CMS	Centers for Medicare and Medicaid Services
CNAMTS	Caisse nationale de l'assurance maladie des travailleurs salariés
CSH	Conseil Supérieur des Hôpitaux
CWA	Civil Works Administration
EBM	evidence-based medicine
FERA	Federal Emergency Relief Administration
GHC	Group Health Cooperative
HAD	l'hospitalisation a domicile
HAS	Haut Authorité de Santé
HCSP	Haut Conseil de la Santé Publique

HPS	Department of Health Promotion and Noncommunicable Disease Prevention and Surveillance
INH	Institut National d'Hygiène
INSERM	Institut National de la Santé et de la Recherche Médicale
InVS	Institut de Veille Sanitaire
LEI	Life Extension Institute
MLI	Metropolitan Life Insurance
MOH	medical officer of health
MRC	Medical Research Council
NCI	National Cancer Institute
NHES	National Health Examination Survey
NHLBI	National Heart, Lung, and Blood Institute
NHS	National Health Service (UK)
NHS	National Health Survey (US)
NIH	National Institute(s) of Health
NYAM	New York Academy of Medicine
OASI	Old Age and Survivors Insurance
OECD	Organisation for Economic Co-operation and Development
PAC	public assistance committee
PCMH	patient-centered medical home
PHC	public health committee
PHLS	Public Health Laboratory Service
PHS	Public Health Service
PMR	physical medicine and readaptation
PPS	prospective payment system
RMP	Regional Medical Program
WHO	World Health Organization
WPA	Works Progress Administration

Chronic Disease
in the Twentieth Century

Introduction

This work has its origins in confusion. In December 2005 I had the good fortune to be invited to a workshop in London devoted to "chronic disease," a term I thought I understood. I left the meeting without the faintest idea what it meant (my own paper was on premenstrual syndrome [PMS], which gives some idea of just how confused I was) and with a simple but nagging question. What holds together a concept that includes a cancer that may kill you within months, diabetes that requires lifelong management, and senility in old age (not to mention PMS, which the meeting organizers at least accepted more or less as a chronic disease)? After an initial period of research I found, somewhat to my surprise, that while a large literature builds on the premise that chronic disease is hugely significant, no one had examined the historical development of the concept in the twentieth century, with the exception of several articles and one policy-oriented book by the historian and former president of the Milbank Trust Daniel Fox.[1]

While several related but distinct categories like incurability, handicap, and disability have been analyzed as evolving constructs,[2] chronic disease has been dealt with as a more or less natural category originating in "epidemiologic transition" and whose existence thus requires little explanation. The sociologist David Armstrong, it is true, has questioned the "natural" status of the concept and has called for a "genealogy" of chronic disease, presumably along Foucaultian lines. But the overwhelming consensus seems to be that contemporary concern with chronic disease is simply the logical consequence of changing disease patterns, notably the apparent conquest of infectious disease that has led to new causes of morbidity and death. Even if this explanation seems plausible from the perspective of the early twenty-first century, I do not believe that it is sufficient. Far too much remains unexplained, as I try to demonstrate throughout this book. Rather than "genealogy" that starts from the present and looks backward, I conceive of this work as "natural history" that starts from the past and works its way forward, which is in fact how I researched and wrote this book.

Because I have been asked so many times what diseases I am examining, I must emphasize that this is a book not about specific diseases, but about a meta-concept that has been exceptionally elastic and fluid. The term *chronic disease* has covered different diseases and conditions depending on the time, place, and goals of those utilizing it. The emotional valence surrounding it and the uses to which it has been put are equally varied. Individual diseases are discussed in this book only to the extent that they intersected with and helped shape the meta-concept or competed with it for funds and public attention. Thus, cancer and diseases of older persons are more prominent herein than is mental illness, which followed an independent trajectory that intersected only sporadically with that of "chronic disease." But just as we have learned to understand many disease concepts to be constructed from a mixture of elements—symptom patterns, biological theories, scientific artifacts like statistical correlations or histological slides, and, occasionally, entrepreneurial ambitions and sociocultural assumptions—"chronic disease" must also be treated as a complex conceptual construction that allowed the world to be organized in particular ways and that was mobilized in varied ways depending on time and place. In other words, the concept has done diverse kinds of work for those using it. The aim of this book is to understand its use or nonuse in different times and places.

Chronic disease is not a new notion. It has been a significant category for understanding illness since the Greeks and Romans. To fully appreciate how its meaning and connotations changed in the twentieth century, we must first understand what it meant and how it was used in earlier periods.

Chronic Disease before the Twentieth Century

Chronic disease has traditionally been understood in contrast to acute disease. Beginning in the Greco-Roman period, it was common for comprehensive medical treatises to divide diseases into categories of acute and chronic. While the former run their course quickly, either killing you or going away (either naturally or due to the physician's intervention), chronic diseases linger. They, too, have a course to run, but one that is much slowed down. If lengthy temporality was the main definer of chronic disease, there was little consensus about what length of time constituted chronicity. Was it forty days or more, an indeterminate lengthy period, or the fact that, over time, a condition never improved or worsened? For some authors of medical works, chronic diseases were more intense and painful were than short, acute illnesses. Aretaeus the Cappadocian (second century AD), for instance, began his discussion of chronic disease as follows:

> Of chronic diseases the pain is great, the period of wasting long, and the recovery uncertain; for they are not dispelled at all, or the diseases relapse upon any slight error; for neither have the patients resolution to persevere to the end; or, if they do persevere, they commit blunders in a prolonged regimen. And if there also be the suffering from a painful system of cure, —of thirst, of hunger, of bitter and harsh medicines, of cutting or burning, —of all which there is sometimes need in protracted diseases, the patients resile as truly preferring even death itself.[3]

For others, in contrast, these diseases were milder and less intense.[4] Some medical thinkers minimized the differences between the two terms because of the arbitrariness of declaring a disease chronic after a certain period of time, and because acute diseases could become chronic and vice versa. The great Paris clinician A. F. Chomel wrote in 1822 that, useful as it undoubtedly was, the distinction should not be granted too much importance.[5] But while the fluidity of categories was generally recognized, many medical authors did in fact view chronicity as a real category. Before the mid-nineteenth century, it was common to think of illnesses as imbalances or derangements of individual constitutions. So it is not surprising that many of the books and articles on chronic disease proposed a single cause for many different chronic conditions. Some of these were very simple indeed. William Cadogan, in his often-reprinted book of the eighteenth century, found that gout and all chronic diseases were the results of the same causes: "indolence, intemperance and vexation." Since individuals brought about such illness through their behavior, the solution was not medicine but "it must be gently calling forth the powers of the body to act for themselves, introducing gradually a little more and more activity, chosen diet, and, above all peace of mind, changing intirely [sic] that course of life which first brought on the disease."[6] A century later, John King attributed most chronic illnesses to "temperament" and poor hygienic practices. Consequently, "although there are many maladies in which medicines and medicinal treatment can not be dispensed with, yet I am fully convinced that nearly, if not quite, one-half of the sicknesses which come under the care of medical men, could and ought to be cured solely by a recourse to hygiene."[7]

Samuel Hahnemann, the founder of homeopathy, had a more complex view of chronic diseases. Most chronic diseases, he thought, were, like acute diseases, the result of original miasms, dynamic noxious influences of which there were three, including one that was the source of syphilis. He called the most important and oldest of the three "Psora," which originated in leprosy, then evolved into less dramatic skin itches and eruptions, and had over time become internalized within the body, causing literally thousands of different illnesses. It had in the last three hundred years become "the most universal mother of chronic diseases." Medical mistreatment had exacerbated the situation by destroying the itch eruptions on the skin that mitigated internal conditions. The only possible cures were one or more of the homeopathic remedies that acted directly on the Psora.[8]

If purportedly common causes could hold together the category of chronicity, so could therapies. Mineral waters were probably the single therapy most closely identified with chronic diseases.[9] Mineral waters had some degree of scientific status in nineteenth-century France (although not in the United States), but other therapeutic solutions for chronic illnesses got less respect. A British review of 1851 had great fun ridiculing a book that saw exercise as the remedy for all chronic diseases.[10] More successful was electricity, often proposed as a therapy for sufferers of many conditions.[11] At the end of the eighteenth century, the British physician Thomas Beddoes collaborated with Humphry Davy to establish a pneumatic institution in Bristol to treat patients with "airs," or gases, produced through chemical reactions. Treatment was initially aimed at tuberculosis and other respiratory conditions, but the opening announcement in the *Bristol Gazette* also welcomed sufferers of "Palsy, Dropsy, Obstinate Venereal Complaints, Scrophula or King's Evil, and other Diseases, which ordinary means have failed to remove."[12] Specific climatic zones and even the blue color of the sky might also have a healing effect on human and animal diseases.[13]

Although much discussion of chronic disease was at this simple level, some medical writers of the early nineteenth century, particularly in France, tried hard to do justice to the variety of conditions represented by the term. They distinguished between functional, localized, and multi-organ chronicity; between those conditions caused by active illnesses and those resulting from organic lesions that remained after the illness had passed; between illnesses caused by a constitutional predisposition (diathesis) and those caused by an external cause; between simple forms and complex forms caused by coexistence of factors or by factors interacting dynamically; between those that were incurable and those that responded to treatment, however slowly. [14] These were not just theoretical questions but ones on which treatment, whether active or palliative, could be based. In 1854 the Paris Academy of Medicine, which tried to resolve controversial medical issues through debate, devoted its longest discussion of the nineteenth century to the question of whether microscopic analysis of cells could identify them as either curable or incurable. If it was a true cancer cell, the microscopists claimed it was incurable, whereas surgeons asserted that some cases of cancer were in fact curable.[15]

Hundreds of books on chronic disease were published in the eighteenth and nineteenth centuries, suggesting that it was far more common than official mortality statistics or historians' focus on infectious disease would suggest. William Osler wrote in his *Principles and Practice of Medicine* that many of those who died of infectious diseases were already gravely ill: "There is truth in the paradoxical statement that persons rarely die of the disease with which they suffer. Secondary terminal infections carry off many patients with incurable disease."[16] Nonetheless, determining the extent of the problem was almost impossible. The historian Gerald Grob has suggested that Americans before 1900 were mainly preoccupied with acute infectious disease although many in fact suffered from chronic ailments.[17] This situation was less a result of lack of interest than of the difficulty of getting doctors or families to talk about such matters. The reason offered in 1925 for the secrecy surrounding cancer was applicable to many chronic diseases during the previous century: "the implication of hereditary taint or wrongdoing or both."[18] In 1880 John Shaw Billings tried to include in the census information about the nature of diseases and disabilities that people suffered; his attempt provoked little controversy but also garnered little useful information. When ten years later he pursued the matter in greater detail, including questions like "whether suffering from acute or chronic disease, with name of disease and length of time afflicted; whether defective in mind, sight, hearing, or speech, or whether crippled, maimed, or deformed, with name of defect,"[19] and then sent a circular to doctors asking that they provide information about the chronic diseases of individual patients, a firestorm of controversy erupted. Lawyers and politicians became involved and doctors refused to participate. (Other questions, notably one asking about mortgages, were equally controversial.)[20] The questions did not reappear on the 1900 census, but a 1905 census in Massachusetts asked about acute and chronic diseases (defined by length regardless of type) and found that 1.12 per 1,000 responders had an acute illness and 7.9 per 1,000 a chronic illness.[21]

While chronic diseases were certainly significant realities for sick people, their doctors, and some compilers of statistics, they were not in the nineteenth century a major category for public health. Even lingering diseases like tuberculosis, the leading cause of mortality, did not become objects of public health activity until the latter decades of the nineteenth century. In part, this resulted from lack of accurate knowledge about their causal mechanisms or prevalence and in part from their endemic nature. They were not, like cholera, plague, and yellow fever, sporadically imported from abroad, dramatically killing large numbers of people in a short space of time. And there were no relatively simple *public* actions—like vaccination, quarantine, or sanitary reform—that were widely believed to be effective. One of the consequences of the bacteriological revolution was to gradually transform tuberculosis from a "reputedly" incurable chronic disease and syphilis from a treatable (using mercury) though shameful chronic disease into potentially curable infectious diseases susceptible to large-scale public health interventions. A recent debate among historians about whether infectious diseases were the dominant health problem of the nineteenth century can be largely explained by the fact that those arguing that infectious disease have been overemphasized are historians of disease while the defender of infectious disease predominance is an historian of public health.[22]

During the course of the nineteenth century, specific disease entities superseded general constitutional disturbances as the chief focus of medical thought. The revival of constitutional medicine in elite circles early in the twentieth century gave new life to comprehensive views of chronic disease. For instance, the well-known Boston orthopedist J. E. Goldthwait wrote numerous articles in eminently respectable medical journals arguing for a mechanistic conception of chronic diseases. In his view, poor posture displaced the internal organs of the body, impeding their function and causing any number of chronic conditions, depending on which organs were affected.[23] Nonetheless, on the whole, a general notion like "chronic disease" became less useful within elite mainstream medicine and became increasingly associated with alternative medicines and new gadgets. Although academic medical writers never abandoned the idea that some diseases were essentially chronic, they more frequently used the words *chronic* and *acute* as adjectives, either of which might be applied to conditions like gastric ulcers, abscesses, communicable diseases, or isolated joints and body parts.[24]

But if the term *chronic disease* lost much of its specific character in the world of medical science, it retained and even increased in importance as a term of administration and welfare, often serving as popular shorthand for more regulatory formulas like "incurable, infirm, or elderly." Assignment to one of these categories might give paupers access to certain types of institutions like hospices as well as meager welfare benefits and/or charity. But such labeling also had serious restrictive effects; as general hospitals gradually lost their welfare function and became more medicalized, they frequently denied admission to "chronics" on the grounds that there was nothing medicine could do in such cases and that hospital beds, always in short supply, should be reserved for those who might benefit from medical treatment. Certainly regulations to this effect were often breached, but they nonetheless caused considerable suffering for many. That "chronics" were taking up space in hospitals was a frequent complaint and how to get rid of them was a constant theme during the nineteenth and early twentieth centuries. But there did exist genuine concern in certain quarters about the best way to care for people suffering from such conditions and this concern stimulated charitable works and welfare initiatives. Such concern, however, spread only sporadically and briefly into the wider political and social arenas.[25]

That rules were frequently not enforced in general hospitals helps explain how physicians could witness and write about diseases like cancer and neurological conditions. J. M. Charcot was able to write *Lessons on Chronic Disease*, the single most famous nineteenth-century medical work on this subject,[26] because the ethos of research and teaching that had begun to sweep European hospitals since early in the nineteenth century had spread to the two huge hospices of Paris. Charcot became senior physician at the Salpêtrière in 1862. This hospice had both a psychiatric service and one for elderly and chronically ill patients. The first provided Charcot with the medical cases that made him a pioneer of modern psychiatry; the second made possible much of his work on chronic diseases, principally neurological diseases like multiple sclerosis and cerebral hemorrhages but also conditions like gout and chronic rheumatism.

Chronic Disease in the Twentieth Century

In the twentieth century, although the term remained imprecise and elastic,[27] *chronic disease* acquired new meaning as one of most serious problems facing national healthcare systems. One has merely to do a Google Scholar Search for "Chronic Disease Care" or read the web pages of national Ministries of Health to see that it has become ubiquitous. The process was of course gradual. Appearing in odd years as a category in *Index Medicus* during most of the 1920s, "Diseases, chronic" became a category that appeared annually from 1927 on. In 1935 the US Public Health Service (PHS) undertook the National Health Survey (NHS) of 1935–36, an unprecedented large-scale survey to determine the incidence of chronic disease and disability. In 1955 a new periodical, the *Journal of Chronic Disease*, made its appearance. The term *chronic disease* was joined by other terms like *chronic illness*, which had rarely been used during the previous century but which became increasingly popular.[28] Although *chronic illness* came to refer to the subjective experience of sickness, at least in social science circles from the 1970s on, initially it was largely synonymous with *chronic disease* but conveyed a softer and more hopeful message, one with fewer associations with poverty and incurability and thus more appropriate for a growing social movement. Another new term used somewhat less frequently was *degenerative disease*, which conveyed the sense that aging led to a natural deterioration of organs. In some usages, it also conveyed a sense of collective and debilitating national decline due to social change and "civilization."

For most historians and sociologists the rising interest in chronic disease throughout the century simply reflects changes in disease patterns in the Western world that took place as a result of the growing capacity in Europe and North America to prevent and cure infectious diseases. Infants who would have died in the nineteenth century lived on into adulthood in the twentieth century. Adults who would have previously died from infectious diseases were saved by sulphonamides and postwar antibiotics and thus moved on to the diseases of middle and old age. By the second half of the twentieth century, many of those suffering from serious noninfectious diseases that could not be cured could survive for increasingly prolonged periods with proper management; and by the end of the century, more and more people were living long enough to deal with multiple health problems of extreme old age. The issue was not just growing numbers of chronically ill people; it was more fundamentally the costs of their care. High-tech acute care is very expensive but time-limited. Long-term care has no such time limits and may be punctuated by costly acute episodes. All this is certainly part of the story of chronic disease in the twentieth century.

Two other approaches add to our understanding. There is a Foucaultian approach that is used by relatively few analysts but that deals directly with the rising concern with chronic illness. It works on a very different analytical level from the epidemiological argument. Whereas D. M. Fox and others see efforts to adapt medical institutions to a new disease reality as largely failing, these writers view concern with chronicity as part of a profound structural change that has in fact been remarkably successful. W. R. Arney and B. J. Bergen perceive developing concern with chronicity as part of a wide-ranging transformation that allowed medicine to organize itself around the concept of prevention and to take over problems that had previously been managed by social welfare institutions providing assistance or charity to certain kinds of indigent people (whom they refer to as "anomalies").[29]

David Armstrong has associated the spread of the notion of chronic disease with the rise of a new kind of medicine he calls "surveillance medicine," replacing the "hospital medicine" that predominated for nearly two centuries; this new regime takes patients out of hospitals and focuses on prevention of a growing number of conditions through ever-more pervasive and sophisticated forms of monitoring and screening, including self-monitoring.[30] One can of course take issue with many aspects of such interpretations. Surveillance and preventive medicine began with infectious diseases like tuberculosis and syphilis, and there is no reason to think they would not have spread as states took a more active role in healthcare, no matter what kinds of diseases were targeted. This reading, moreover, ignores the specific historical mechanisms that link chronic disease and surveillance. And reports about the death of hospital medicine have, to paraphrase Mark Twain, been greatly exaggerated. What we have seen rather is enormous economic pressures to do what hospitals do more quickly and cheaply and sometimes in alternative settings. Still, with all these caveats in mind, something like "surveillance" was very much part of the effort to confront chronic disease. Most recently, Armstrong has argued that emphasis on chronic disease has largely been about pathologizing the process of aging, implying both the possibility and the necessity of treatment under medical control.[31] While this argument neglects the quite different motives of many groups advancing a chronic disease agenda and underestimates the difficulties that aging posed for this agenda, Armstrong correctly points to a major cultural shift: emphasis on chronic disease has been about rejecting attitudes of hopelessness, inevitability, and neglect applied to a variety of conditions, including those of old people, and transforming them into targets of intervention and amelioration.

In making his latest arguments, Armstrong exemplifies a third approach that analyzes the "medicalization" process through which more and more conditions are defined as diseases. If it rarely focuses on the broad category *chronic disease*, this approach may center attention on particularly susceptible population groups like women or older people. Most often, however, the units of analysis are the new conditions that, according to successive waves of interpretation, were promoted by physicians or medical scientists, then by pharmaceutical companies, and, in more recent work, by patient groups;[32] these are frequently long-term conditions like attention deficit hyperactivity disorder (ADHD) or fibromyalgia. Not only do new chronic diseases appear with considerable regularity, but older conditions also expand. Historians like Robert Aronowitz and Ilana Löwy have described the bracket creep that defines diseases like cancer at earlier and earlier stages.[33] Similarly, criteria for defining hypertension (a risk factor that has become a disease), autism, or depression tend to become more and more inclusive. Much of this literature is highly critical of this process. But there is little question that such developments inflate the number of people defined as chronically ill and requiring treatment.

While there is truth in all of these approaches that together explain a great deal that occurred—in combination with the process of medical specialization early in the twentieth century that created groups of cancer or heart specialists who founded such disease-based associations as the American Cancer Society or the American Heart Association that gave new prominence to the diseases they treated—they nonetheless leave a number of critical questions unanswered.

The first of these has to do with timing. As I demonstrate in the first few chapters, discussion of chronic disease began during World War I and intensified during the 1920s and 1930s. Although infant mortality had significantly decreased and mortality from many infectious diseases was declining, tuberculosis remained one of the great killers. Mortality rates overall were going down so it is hard to understand why the apparent decline of some diseases that killed the young and the rise of others later in life inspired so much passion. Furthermore, there was no unanimity about the reliability of statistics showing a dramatic increase in mortality from cancer and heart disease; in fact, many observers, including the producers of these statistics, were highly skeptical about them on the grounds that disease categories, modes of diagnoses, methods of reporting, and medical fashion had changed drastically. At the very least, one has to look carefully at who took such statistics seriously and why they did so.

Second, even if one accepts some rise in the incidence of diseases of middle and old age—and everyone admitted that declines in infant and child mortality inevitably led to increases—why jump from specific diseases to a general category like "chronic disease"? The statisticians who produced the data certainly did not do this. If one looks at the early statistics produced, one sees classification based on organic systems, not on acuteness versus chronicity. One can certainly understand that claims of a rising incidence of cancer, whether correct or not, might provoke widespread terror. But how does one explain the influence of a fairly abstract meta-concept like "chronic disease"? One answer is that new types of quantification objectified the category by counting such new "facts" as "days of disability" or "risk factors" and later "quality of life." But one must still ask why and how such categories were created. Another response is that this was part of nascent processes of *medicalization* or surveillance, which in hindsight is undoubtedly true but which does not get us far in understanding why the process was occurring at that particular time and place.

The word *place* leads to the third and most critical point for the purposes of this book: large-scale public and political concern with chronic disease was largely an American phenomenon until roughly 1960, when it began to spread to other nations. The reason for this American interest is that a variety of groups found it a useful category for pursuing their quite distinct goals and eventually for co-alescing around a policy of healthcare expansion. In the United Kingdom, chronic disease appeared as a social issue just after World War II, referring to the problem of elderly patients either in grossly inadequate Poor Law hospitals or "blocking" beds in general hospitals. By the 1960s, the broader American understanding of chronic disease came into use in the United Kingdom but was not central to Brit-ish discussions of healthcare for much of the twentieth century. In France the term was also applied primarily to elderly persons and was used to describe patients whom administrators wished to evict from hospitals. The term in fact disappeared from official discourse after 1970 because it was thought to be stigmatizing and because other terms were more congruent with administrative thinking about the provision of healthcare facilities. Such differences raise several obvious ques-tions that will be explored in this book. What conditions and forces in the United States made "chronic disease" such a potent way of thinking about and organizing the world of healthcare? How did other nations understand and deal with the phenomena that "chronic disease" covered? And finally how does a meta-concept like "chronic disease" move from one environment to another?

To avoid misunderstanding, let me emphasize two points. I am *not* suggesting that other nations were uninterested in diseases like cancer or heart disease. One could in fact make the case that the United Kingdom and France were well in advance of the United States in providing state support for cancer care. Nor am I suggesting that the notion of chronic disease was completely absent from rheto-ric and policy in the United Kingdom and France. What I am arguing is that the concept of chronic disease was considerably more central to healthcare and wel-fare rhetoric, organization, and policy in the United States than elsewhere, at least until the 1960s and probably well beyond. The second point that I wish to stress is that I am not seeking to enter debates about American exceptionalism. If anything, I believe all nations to be exceptional. The French story I tell in chap-ter 10 is surely as unique as anything that occurred in the United States. If I had chosen another meta-concept like "exclusion" or "handicap," my narrative would have centered on Europe.

The account I offer is about more than differences among countries. It is about teasing out the various components that make up the notion of "chronic disease." As elastic as this notion has been with respect to the illnesses that it covers, concern with chronic disease has been of three basic types. The first is medical *treatment* of the sick, which was indeed a problem in the early twentieth century when few facilities for patients who had cancer, heart disease, and other conditions existed. My conception of medical includes research because it was recognized that treatment rested on both biological research to explicate disease mechanisms and clinical research to test therapies. A second element in chronic disease discourse and practice is *prevention*, the mission of public health. From the earliest days of the chronic disease movement, there was concern to prevent conditions that killed, maimed, or disabled. By the middle of the twentieth century, this form of prevention was considered "primary" because a second form of prevention had appeared on the scene: finding and treating existing disease in its early stages in order to prevent deterioration and possible death. Doing so, of course, required medical care and brought the notion of prevention and public health policy more directly into the medical sphere. And finally there is long-term *care*—a term that includes medical care but much more as well. Very ill and infirm people frequently require nonmedical services and facilities. One of the thorniest issues surrounding chronic illness is how to provide such services. The task is complicated by the tendency to place nonmedical support under the jurisdiction of welfare authorities. The division of labor between medical and welfare institutions has frequently been a point of tension and conflict in all three nations under discussion not because of imperialistic ambitions on either side but because both wished to *avoid* responsibility for such services.

Criticism of chronic disease policy usually focuses on how money and resources have been distributed among these three components. Characteristically, it has been charged that too much weight has been placed on treatment (therapeutics and research) and not enough on prevention or care. Alternatively, criticism focuses on particular strategies within one of these elements; that medical care, for instance, focuses too much on acute episodes of chronic illness and not enough on proper disease management to avoid these episodes. Primary prevention is short-changed by emphasis on secondary prevention and may itself be inadequate because of a dominant focus on individual behavior modification that ignores social factors like poverty and unhealthy environments. Long-term care, it is alleged, invests too much on institutionalization and not enough on enabling people to live at home. However justified such criticisms may be, one cannot say that chronic disease is not taken seriously. Disagreements are not about the seriousness of the problem but about the best way to deal with it.

Separating the different elements that make up "chronic disease" is not the only analytical distinction that is required. There is first of all the distinction between movements or programs on behalf of specific diseases like cancer or specific populations like older people and those seeking more comprehensive solutions based on the assumption that all such problems have fundamental similarities and require common or at least coordinated solutions. Specific diseases may have strong emotional resonance because people suffer from them directly; the idea of chronic disease functions at a more abstract level. Practical programs based on specific diseases are simpler to manage than comprehensive programs but they may lead to waste and gaps in service. The strength of the cancer movement has been at times a major asset for promoting the comprehensive view of disease (see chapter 2), but like other disease-specific movements has also competed for resources, often successfully, against comprehensive chronic disease strategies. The organization of the National Institutes of Health (NIH) around specific disease categories is perhaps the most striking example of this dynamic.

There is a second distinction that must be raised. Some problems occupy small groups who administer healthcare institutions and who make decisions that provoke relatively little public response outside narrow professional circles. Other issues enter the public arena and generate widespread and passionate debate among many interested groups and actors. The narrative of this book attempts to account for the early move in the United States from the first to the second type of social action and the later, considerably less charged, and differently tilted trajectories followed in the United Kingdom and France. This distinction is particularly significant because it underlies my argument that concern with and responses to chronic disease involved political choices that are not directly addressed by either the epidemiologic transition or medical surveillance arguments. That chronic disease was the major problem of the twentieth century was an idea embraced by many groups for a wide variety of reasons: administering and planning for large populations; improving the quality of the race; attempting to expand the functions and raise the status of organizations or occupational groups; promoting healthcare reform or on the contrary resisting such reform; increasing investment in biomedical research; attempting to control healthcare costs; and, occasionally, alleviating suffering.

PART I / Chronic Disease in the United States

CHAPTER ONE

"National Vitality" and Physical Examination

In the early twentieth century, Americans approached their health, both individual and collective, with a mixture of confidence and anxiety. Confidence was based on the sharp decline of infant mortality; increasing control over infectious diseases; and the promise that science and rational management, backed by American initiative and money, would lead to ever greater health and progress. Anxiety resulted from a widespread feeling that Americans were in the throes of collective physical decline and from the growing visibility of frightening non-infectious diseases like cancer, heart disease, and diabetes. Mortality figures that suggested a sharp rise in such diseases were highly controversial but seemed convincing to many because of the many dramatic transformations rocking American society.

Urbanization and civilization generally, it was believed, had transformed human life in unnatural ways that damaged health. For many, immigration and race mixing, as well as public health measures that permitted the survival of the unfit, were leading to racial deterioration. The rise of new specialties, like cardiology and oncology, with their associated societies and journals, brought their diseases to the notice of physicians and laypeople. Specialists played a key role in creating the American Society for the Control of Cancer in 1913 (which became the American Cancer Society), the American Heart Association in 1924, and the American Committee for the Control of Rheumatism in 1928. These publicized the terrible toll that such diseases were taking and offered the possibility of cure if enough energy and money were invested in the task. New therapeutic modalities like occupational therapy, physical and rehabilitation medicine, and radiation therapy for cancer made these diseases appear less hopeless, while periodic revisions of the International Classification of Disease made them ever more prominent. By the second decade of the twentieth century, physicians utilized a very different knowledge base than their predecessors had done when filling out death certificates.

But this increasing prominence applied to specific diseases rather than to an entire category called "chronic disease." One relatively modest movement did much to set the stage for the American preoccupation with this broader category. This was the alliance supporting periodic medical examinations. It identified central issues and generated vocabulary and clichés for talking about them that would persist for decades. It was closely connected with the insurance industry but extended far beyond it, exerting considerable influence on public health agencies and associations of physicians and hospitals. Some of its leading figures were supporters of eugenics, which had emerged from the larger Progressive movement. In this arena they did not emphasize reproductive strategies for improving the race, but rather the Progressive assumption that the health of children and adults could be enhanced through the application of scientific knowledge.[1]

Periodic Examinations and the Insurance Industry

In 1908 President Theodore Roosevelt, having recently created a commission to examine conservation of forests, mineral resources, and water, added to it the eminent Yale economist and supporter of eugenics Irving Fisher to report on health conservation. His *Report on National Vitality* estimated that annual monetary loss from preventable deaths amounted to roughly $1 billion, based on potential earnings, with illness responsible for another half-billion-dollar loss.[2] He claimed that it was possible to significantly reduce this waste of life by simply acting on existing knowledge about many diseases. Attributing much of the blame for illness to civilization, he asserted that balance could be restored through teaching correct personal hygiene and encouraging periodic medical examinations to find and remedy "defects" and modify behavior.[3] His views reflected the general quest for efficiency, order, and scientific management that characterized the early decades of the century, as well as the eugenic views popular at the time.[4] Similar considerations led to the rapid expansion of medical inspection of schoolchildren during this period. Starting in Boston in 1894, periodic school medical inspection was by 1914 mandatory in seven states and permitted in thirteen more.[5]

The idea of periodic physical examinations can be traced back to the nineteenth century. But Fisher's championship of them had greater salience by the early twentieth century, particularly for the life insurance industry, which had a financial interest in keeping plan holders alive as long as possible. Industrial life insurance companies like Metropolitan Life (MLI), Prudential, and John Hancock offered policies to the working classes for small weekly premiums that were often collected door to door. Lee K. Frankel and Louis Dublin, key figures at MLI, were active in the public health movement. Frankel came from the world of welfare care for Jewish immigrants. As head of the MLI welfare committee, he introduced such innovations as the collection of morbidity statistics about policyholders. Early in the century, MLI and other insurance companies pioneered in utilizing new public relations and advertising techniques to promote healthy behavior. Company agents provided clients with a series of pamphlets on better living. Among those who wrote such pamphlets for MLI were Irving Fisher and noted public health figures like Milton Rosenau and Charles Winslow.[6] These efforts duplicated those of other welfare, public health, and charitable agencies dealing with the poor. The same is true of the visiting nurse services that certain companies began to offer during these years. In 1909, when six hundred organizations sponsored the work of visiting nurses, MLI established an insurance payment scheme for home nursing care. As with health education campaigns, the goal was to prevent premature death. In 1911 MLI extended the program across the country, contracting with visiting nurse associations whenever feasible. The program proved so popular that in the mid-1920s the John Hancock Company, West Coast Life, Travelers, and Aetna Life established nursing services as well.[7]

Neither educational nor home nursing programs focused on chronic illness (including tuberculosis), which constituted only about 9% of MLI cases. MLI, in fact, following a familiar business model, tried to limit home nursing care to acute illness by restricting numbers of visits. But the spread of these nursing programs meant that visiting nurses were seeing more chronic cases. Whether this reflected an increasing incidence or the growing visibility of chronic disease as programs spread is difficult to ascertain. Sophie Nelson, director of the John Hancock nursing service, was commissioned by Boston's Visiting Nurse Association to determine whether home nursing care for chronically ill people was effective and economically viable. She compared the outcomes of unlimited care by visiting nurses to outcomes of care restricted to six visits and concluded that results were the same. While there might be humanitarian reasons for more frequent homecare of the chronically ill, insurance companies were in no position to bear the cost. At most they could pay for very limited care.[8]

Perhaps the most influential aspect of insurance practice was the effort to popularize the idea that everyone should receive regular physical examinations from doctors. In 1909 the Postal Life Insurance Company began a program of periodic physical examinations for policyholders. In 1913 Fisher became a founding director of the Life Extension Institute (LEI), a business enterprise run by a Manhattan contractor and builder named Harold Alexander Ley. The LEI performed physical examinations for various clients but notably on behalf of insurance companies. In Fisher's words, "The chief objective was to harness up the profit making motive of life insurance companies with the task of lengthening human life."[9] By the mid-1920s, more than forty insurance companies utilized the LEI to examine their clients.[10] In many ways the LEI was a victim of its own success because examination on an industrial level made physicians nervous. A lawsuit brought before the Supreme Court of New York was settled in 1935 when the LEI agreed to end all activities that the state medical association deemed medical practice. At the time of the LEI's demise, *Time Magazine* presented it as a thriving

> $2,000,000 concern, occupying three floors of a midtown Manhattan building and offices in Chicago. Its doctors had made 1,620,000 medical examinations during the preceding 22 years. Three of every 100 examinees came on their own initiative attracted by the advertising which the Life Extension Institute no longer finds necessary or by some of 2,000,000 educational leaflets distributed each year. Two were employees whom business concerns needed to keep healthy. The other 95 were holders of insurance policies whom insurance companies wanted to keep alive as long as possible.[11]

The LEI reflected the new optimistic mood of the insurance industry. Leading industry figures saw insurance as a progressive force in American life and as an ally of the public health movement; companies conducted local health surveys and worked directly with public health officials. In 1917 Lee K. Frankel, then third vice president of MLI and soon to become president of the American Public Health Association (APHA), divided the history of insurance (like a good follower of the positivism of August Comte) into three historical phases. The first was one of fatalism. The second period of passivity was characterized by acceptance of mortality rates and premiums based on these rates. The third "positive period" had its origins in preventive medicine. "What we are striving for today is the fulfillment of Professor Fisher's belief that the average length of life may be extended fifteen years by control of preventable diseases. The modern conception of life insurance has added the new function of life extension."[12] The key was to influence the health behavior of clients. There were many ways that insurance companies could do this: through early meetings with agents, an initial medical examination, providing educational advice when premiums were collected, nurse visits, and periodic medical examinations. MLI was one of the LEI's most important clients. In discussing these issues, Frankel never once mentioned chronic diseases. Virtually any condition, he thought, would benefit from periodic examinations. Nonetheless, leaders of the LEI increasingly attached special importance to what they called organic or degenerative diseases.

The LEI's two leading spokesmen, aside from Fisher, came out of the insurance industry: E.E. Rittenhouse had worked for the Equitable and Provident Companies and Eugene Lyman Fisk had been chief physician of Postal Life. But their ideas reverberated in public health, medical, and political circles. These leaders of the LEI argued that the American nation was experiencing a crisis of "vitality," a term that would become something of a cliché in these and coming decades. As early as 1915, Fisk could invoke data from physical examinations to show that Americans were not in good shape. Only 2.4% of insurance policyholders, he claimed, were normal. All the rest were imperfect and needed advice regarding their physical condition or living habits. He reported that about 93% had not been aware of their impairments and 66% had been referred to a physician for treatment. Results were similar for employees sent for examinations by banks and other businesses. Fiske's most striking data concerned the number of young men exhibiting arteriosclerosis and thickening arteries, which promised even higher levels of cardiovascular troubles in later life.[13] As time went on, the LEI generated data about clients that was unprecedented in scope. It constituted a goldmine for the emerging field of morbidity epidemiology. In 1929 the research division of the Milbank Memorial Fund, with the assistance of the US Public Health Service (PHS), made a statistical analysis of more than one hundred thousand coded records of the institute's examinations. This resulted in a series of articles published in early 1930s by Edgar Sydenstricker and Rollo Britten, who would become central to the chronic disease story in 1935.[14] Fisher and the LEI popularized a vocabulary—national vitality, life extension—that would be pervasive for several decades. Three of their ideas were particularly salient for the emergence of the chronic disease movement. First, much disability was preventable; second, periodic medical examinations were the most effective means of prevention; third, many diseases could be averted in this way, but degenerative diseases were particularly amenable to such intervention.

That much illness was preventable seemed to be confirmed by the growing control of infectious diseases and by numerous recent discoveries about human physiology. That most of Fisher's early predictions for reducing death rates were in fact being achieved or surpassed by the 1920s seemed to further corroborate his acumen. That periodic medical examinations constituted the best path toward further improvement was more difficult to demonstrate but seemed plausible. The campaigns of the National Tuberculosis Association to find and treat the illness at an early stage seemed at least partly responsible for declining tuberculosis mortality rates and served as a model for the movement to combat cancer. But overall, proponents of periodic examinations based their views on the commonsense notion that many conditions that could be successfully treated at an early stage led to death or incapacity when left untreated. Furthermore, periodic examinations gave physicians the opportunity to impart basic health information to patients. As time went on, more and more conditions really could be treated at an early stage; syphilis, diabetes, and infectious diseases like rheumatic fever that could cause lifelong debility. Many at the time believed that "focal infections" of the teeth, tonsils, intestines, and other organs caused a variety of serious conditions. By the 1930s, the development of nutritional science gave the LEI and other advocates of prevention increased capacity to offer medical and hygienic advice based on apparently solid scientific knowledge.

Many treatable conditions, however, were in fact highly symptomatic and one could expect anyone afflicted with them to seek medical help if they could afford it. What then justified periodic examinations? Here the notion of "chronic" or "degenerative" diseases came into play. Fisher's original report had little to say about this subject primarily because there existed little knowledge about prevention that he could utilize. But others had strong views on the subject. Fisher's collaborator in both the LEI and the more visibly eugenic Race Betterment Foundation, J. R. Kellogg of cereal fame, pointed out that many supporters of eugenics believed that the white race was suffering from race degeneration, manifested in higher rates of degenerative and chronic diseases like arteriosclerosis, diabetes and heart disease, and insanity and idiocy. The reason was preservation of the weak and unfit and the shift to an unhealthy indoor life. In 1912 William Schieffelin, chairman of the Executive Committee of One Hundred on National Health that had worked closely with Fisher in preparing his *Report on National Vitality* and himself a prominent social worker, justified the need to create a National Bureau of Health, on the grounds that, despite the decline of deaths from acute diseases, mortality from chronic ailments had increased substantially. He discussed explanations that had been advanced by reputable scientists, blaming too much protein in diets and nitrogen famine. He himself attributed "race degeneracy" to many causes, including alcohol, tobacco, assimilation of foreign people of the working class, the strains of the urban environment, and indoor life.[15] In

1910 E. E. Rittenhouse, who would later go on to work at the LEI, emphasized the increased waste from noncommunicable diseases "due to the early wearing out of vital organs due to excesses in eating, drinking, working and playing—in short, intemperate living and the strenuous life." He proposed that the state introduce permanent education campaigns and free periodic medical examinations "to prevent the waste of life and energy which has been going on unchecked."[16]

Giving special meaning to such arguments was the apparently rising death rate among men over 40 (or 55 in some accounts) and mortality rates that seemed considerably worse than those of Britain and Europe.[17] Such fears of population degeneration due to rapid social change, "civilization," and/or "race-mixing" were hardly unique to Americans, and public health figures in many European countries were advocating new strategies of prevention. But what may have distinguished Americans was exceptional optimism and hope. In 1915 Fisher and his LEI collaborator, Eugene Lyman Fisk, published a self-help book called *How to Live*, with a foreword by former President William Howard Taft. Taft suggested that although communicable diseases remained a serious threat, "warfare against such maladies [was] well organized" and it was time to move on to chronic "preventable" diseases "that sap the vitality of the individual and impair the efficiency of the race." The goal of this book, according to the ex-president, was to teach people to preserve their health, improve their physical condition, and increase vitality. "It aims to include every practical procedure that, according to the present state of our knowledge, an athlete needs in order to make himself superbly 'fit,' or that a mental worker needs in order to keep his wits sharpened to a razor-edge."[18] And Fisher expressed his own evolutionary optimism: "A great health movement is sweeping over the entire world . . . Science, which has revolutionized every other field of human endeavor, is at last revolutionizing the field of health conservation." The United States, he suggested, had a great deal to learn from Europe, where individual hygiene was more generally applied and where death rates were declining for all age groups (both highly arguable points).[19]

Beyond Insurance: Public Health and Chronic Disease in New York City

The insurance industry had a close relationship with New York City's Health Department, known for its pioneering innovations. In 1892 the department established the first municipal bacteriological laboratory in the United States, providing free diagnostic services to physicians. That year, it also introduced a program of tuberculosis control. In 1895 it established the first North American laboratory to manufacture diphtheria antitoxin and a year later began requiring permits for the sale of milk. Innovation continued under Herman Biggs, general medical officer from 1901 to 1914. In 1908 a Child Hygiene Division was organized and in 1912 a program of venereal disease control was introduced. The activism of Biggs's department (not to mention a profound aspect of American culture) was expressed in its motto: "public health is purchasable." Biggs had written as early as 1904 that public health in the United States was excessively narrow in scope and "out of harmony with the spirit and development of modern scientific medicine." Statistics seemed to demonstrate that acute respiratory diseases, cancer, and diseases of the circulatory system and of the kidneys were the only major causes of death that had increased during the last twenty years. He himself was convinced that this was not simply due to greater accuracy in reporting but rather reflected a critical sanitary problem that health authorities had yet to confront. He asked with genuine curiosity: "What are the factors in the lives of the inhabitants of our large cities (for I have found that similar increases, less extensive, however, have taken place in London, Paris and Berlin) which have caused such a remarkable increase in the prevalence of these affections, and how are these factors to be removed?"[20] By 1912, the annual report of his department indicated that "the reduction of mortality from the diseases of middle and old age" must be the focus of public health in the future.[21] He would repeat this idea one year later in a talk to the annual meeting of the National Tuberculosis Association.

Soon after, the LEI was established and its leaders publicly supported Biggs's efforts. E. E. Rittenhouse told the Finance Forum of New York City that much life was wasted because the city did not spend enough on the public health department's preventive activities. Provocatively, he characterized the result of the city's "deadly economy" as nothing less than a "communal slaughter."[22] In turn, Biggs by 1923 considered a key public health measure to be physical examination combined with education in healthful living and "the treatment of diseased conditions or abnormalities found."[23] He directly related this strategy to lowering mortality from cardiovascular diseases and the other diseases of later life and called for it to be part of the free services to the poor provided by public authorities. Biggs was not the only influential figure with such ideas. Soon after he succeeded Biggs as head of the city's department of health, S. S. Goldwater sent a brief letter to Dr. Charles L. Dana, chairman of the New York Academy of Medicine's (NYAM) Public Health, Hospital, and Budget Committee. He pointed out that while mortality from "preventable" diseases was decreasing, the opposite was the case for chronic diseases. "Has the time arrived for the formulation of a public health policy that might possibly influence in a favorable way the rate of morbidity and of mortality from those diseases which, according to all the evidence, are not now under satisfactory control? Will you not be kind enough to consider the desirability and the feasibility of arranging, under the auspices of your committee, a symposium before the Academy on this subject?"[24] His goal was to get ideas that could lead to "some useful amplification or modification of the work of the Department of Health."

A subcommittee of the NYAM was struck and reported several months later on the "so called non-preventable diseases."[25] The subcommittee expressed pleasure that the commissioner of health was giving thought to controlling non-preventable diseases, including cancer; pneumonia; and renal, cardiac, and vascular diseases. It nonetheless concluded that attacking them directly was not possible because adequate knowledge of their causes was lacking. Consequently, the subcommittee considered that a symposium would not help the health department to control these diseases. Ignorance, however, did not prevent the report's authors from conveying their personal views on the etiology and prevention. Among the causes of cardiovascular diseases "are mainly various poisons." These could be introduced from without or within by faulty metabolism. Cold and chilling of the body, especially in those whose occupations involved "sudden transitions in surrounding temperatures" (such as firemen and furnace men), were important contributing factors. Consequently, attempts to control these diseases must include "supervision of all the industries and occupations which deal with the manufacture and use of dyes, poisonous metals, drugs, etc." Workers dealing with such materials had to be taught about personal cleanliness as well as about suitable clothing to protect against temperature changes. They should also be taught about "the dangers of excessive nitrogenous foods" and the evils arising from the immoderate use of stimulants, including alcohol, tea, and coffee. The report warned as well of "injurious effects of chronic constipation," unnecessary strain, either nervous or physical, and the possibility of injury to the kidneys following infections, especially those of the tonsils.[26]

The Committee of Public Heath, Hospitals, and Budget of the NYAM recommended to Goldwater that the department of health direct an active campaign to instruct the public in matters of personal hygiene by means of pamphlets, lectures, and the like. A Board of Public Health Education within the department was created in May 1914.[27] The *New York Times* reported on several occasions that Goldwater was calling for the periodic physical examination of every inhabitant of New York City, as well as instruction in "elemental" hygiene. Utilizing Irving

Fisher's technique of promising added years of life, Goldwater claimed that instituting these measures could add an average of three to five years to the lives of New Yorkers. The health department itself had recently inaugurated the practice of physical examinations of its workers and "was adjusting the duties of its employees according to their strengths or weakness. One instance of this was in the case of employees showing symptoms of heart trouble. They are assigned tasks requiring little physical exertion."[28]

In the following months Goldwater continued to be concerned about the health of city employees and residents. He and a group of police surgeons prepared a book, *How to Keep Well*, which was distributed to all policemen. They were advised to be careful about what they ate and drank, and to get plenty of vigorous exercise. Shortly thereafter, Goldwater declared a "War against Drink."[29] In the fall of 1915 he resigned his post to become director of Mount Sinai Hospital. Haven Emerson of the Columbia University School of Public Health replaced him and continued on the path set by his predecessors. Soon after taking office, he called on prominent cardiologists to advise the health department on how best to deal with heart disease. This led to the organization of the New York Committee for the Study and Prevention of Heart Disease, which later became the American Heart Association. Emerson organized a series of demonstrations for physicians on the technique of periodic health examinations and continued Goldwater's efforts to combat alcoholism through health education.[30]

Preventing Death among Productive Men

New York City was unique among American urban centers in taking such vigorous action against degenerative diseases so early in the twentieth century. But it was not completely alone. Rittenhouse reported in 1915 that the New York State Health Department had begun to show interest in this issue (hardly surprising since Biggs was now heading it) and "a part of the state traveling exhibit is devoted to degenerative diseases." He also reported that the US Public Health Service had recently issued an educational pamphlet on exercise and health, "in which attention is called to the lowered expectancy of life above age 40 and to the need of combating the increasing diseases of degeneration."[31] That same year the National Tuberculosis Association called for a week devoted to general physical examinations for everyone.[32]

The NYAM maintained its interest in "non-preventable" diseases and was actively planning a study of mortality from degenerative diseases organized by its executive secretary, E. H. Lewinski-Corwin. The goal was to demonstrate to the public, the medical profession, and public health authorities that despite the general decline in mortality, the death rate of people over 40 had in fact increased as a result of the rising incidence of degenerative diseases.[33] Despite problems deciding what conditions or causes of death should be part of the survey (were industrial diseases, for instance, to be included?), the results of the study appeared in a feature article in The *New York Times Sunday Magazine* in the form of an interview with Lewinski-Corwin.[34] The theme was similar to those popularized by the LEI: "the great waste of human life and national efficiency that is taking place in middle life." The survey had found that the death rate of those over 40 had increased since 1880 by 21.2% in Massachusetts and New Jersey, two states with adequate data. Worse, the increase would continue since deaths from degenerative maladies were increasing yearly. This was tragic because "a man is at his best between the ages of forty and sixty-five." Lewinski-Corwin cited as examples numerous famous men to underline this point. "An ever increasing death rate at forty and over is a tremendously serious subject for mankind to ponder." He summarized and critiqued existing explanations for this phenomenon. First, he dismissed the eugenicist claim that "child-saving" public health campaigns had added many weaklings to the population, thereby reducing its vitality. Since preventive measures were of recent date, they could not have affected anyone over 40. Furthermore, critics did not take note of the fact that many childhood fevers do not kill but rather cause permanent impairments so that preventing them in children reduces rather than increases the incidence of chronic and degenerative conditions later in life. Lewinski-Corwin found insufficient a theory blaming childhood infections for the current rise in death rates since there was no reason to assume that these had increased in quantity. He thus preferred another explanation centered on the environment: "Ill adaptation to a changing, exacting, and unwholesome environment, plus the unhygienic habits and faulty nutrition of the masses are responsible for the early ignorance, intemperance, and immorality—all and every one of them contribute toward health decadence and life waste." Lewinski-Corwin focused on the need to combat alcoholism and prostitution and to educate the public in the principles of hygiene such as more freely using soap and water. It is true that as the mortality from diseases of childhood and adolescence declined one would expect more deaths at later ages. But this rise was not occurring in old age but rather during the most productive period of life, "and this is a most scathing arraignment of our modern nerve and health wrecking civilization."[35]

This preoccupation with the deaths of "productive" men between 40 and 65 (usually expressed *by* men between 40 and 65) continued. But the LEI's statistics based on medical examinations also showed extensive impairments among younger men.[36] Such data was particularly disturbing as Americans by 1915 were contemplating the prospect of entering World War I. "How much longer may we hope successfully to meet the struggles of peace and war with the proportion of inactive, flabby-muscle, low-powered Americans constantly increasing? How long can the nation endure with the physical fitness of its producers and defenders steadily declining?"[37] This situation was all the more dangerous because it was claimed that Europe had not experienced a corresponding increase in either the degenerative diseases or the general death rate in middle life. The reason was simple; social change in the United States had been unusually radical and abrupt. "It is a matter of common knowledge that the high nervous tension under which Americans work and live is virtually unknown in other countries."[38] E. E. Rittenhouse advocated the creation of a "national vitality commission" to seek ways to remedy the situation.[39]

Questioning the Data

Common to all arguments about a crisis of health, particularly among men over the age of 40, was the use of mortality statistics. Not everyone agreed with such readings of the statistics. W. F. Wilcox, professor of economics and statistics at Cornell University, worried more about lowered birth rates due to voluntary personal choice than about lack of vitality.[40] Haven Emerson pointed out how inaccurate American statistics were due to "the lack of thorough clinical and pathological training of physicians" filling out death certificates. Many of the leading causes of death listed on such certificates represented physicians' "unverified impressions" and were "unacceptable"; these included cancer and especially "diseases of the heart and pulmonary emphysema." Of nearly two hundred disease categories commonly used in statistics, Emerson asserted that twenty-*three* should never be accepted without postmortem confirmation and another fifty-three required supporting data. The questionable nature of claims about "diseases of degeneration affecting the age groups over 40 requires us to scrutinize our basic facts before determining the need or character of the campaign for prevention which is urged by those who rest satisfied with the clinical diagnoses of today."[41]

Emerson did not directly attack the claim that "adult" mortality was increasing but his data implicitly called it into question. Emerson would later write: "A few years ago we were quite unnecessarily frightened into the erroneous belief that there had actually occurred a retrogression in the health of adults over forty-five years of age. A more mature, longer and less biased study shows that every age shares to some extent in our advancing safety of life."[42] Despite his skepticism about the rise of degenerative diseases, Emerson played a major role in the development and popularization of physical examinations, arranging demonstrations for doctors while commissioner of public health of New York City and later writing the American Medical Association (AMA) manuals on physical examination. Foreigners equally concerned with chronic illnesses shared Emerson's skepticism about mortality statistics. In 1914 E. F. Bashford, director of the Imperial Cancer Research Fund in the United Kingdom, wrote a sharp critique in the *Lancet* of American figures like Frederick Ludwig Hoffman who were arguing that a dramatic rise in cancer mortality was occurring.[43]

Certainly the most prolific and intrepid questioner of such statistics was Louis I. Dublin, statistician and a vice president of the Metropolitan Life-Insurance Company. While concerned with all aspects of the chronic disease problem, he was for much of his life skeptical of the figures trotted out to dramatize it. In 1913 Dublin wrote his first article on the subject.[44] His starting point was Fisher's by-now famous report on national vitality. Dublin agreed that many causes of death at higher age groups were preventable and thought Fisher's estimates of possible additions to lifespan to be conservative. He also agreed that organic and degenerative diseases like cancer, diabetes, and heart disease accounted for more than half of all deaths; he, like many others, attributed these to the sequelae of acute diseases in childhood and early adult life; habits and behaviors (like alcohol and sexual activity leading to venereal disease); and effects of occupational work among the industrial classes, which he considered particularly significant. Like others, he thought it was possible for those afflicted with functional disturbances to prolong life if directed to occupations that did not "tax their limited energies." Such policies, he pointed out, would lead to increasing mortality at older ages but this was "quite normal and involved no social loss." Dublin's prescriptions for improving matters were largely conventional. But he did emphasize occupational hazard and insisted on better labor legislation, "the improvement of factory sanitation, the medical examination of employees and the instruction of both employers and employees in industrial hygiene." He added: "It will be necessary to supplement labor legislation with the careful examination of death certificates to see that in every instance those who are responsible for preventable deaths are properly prosecuted."[45]

Several years later, Dublin presented another argument based on a modified understanding of eugenics. Increasing mortality at higher ages, he suggested, was

largely due to certain immigrant groups. Mortality of the native-born had remained stable since 1900, when statistics became somewhat reliable. Likewise, not
all foreigners were equal. Jews and Italians had relatively low rates of mortality.
Germans and the Irish had particularly high rates, mainly due to tuberculosis and
degenerative diseases. This in no way diminished the importance of campaigns
for life extension, but it suggested the need for a special effort to "instruct the foreign population in the principles of personal and civic hygiene."[46] A more vigorously eugenic version of this argument focusing on the variation of disease among
"races" was produced around the same time by biologist Charles Davenport.[47]

By 1917, Dublin's qualified reservations had turned into a comprehensive critique of the LEI's driving assumption that mortality of the middle-aged was rising. In the first place, the increases in mortality for men over 45 were very small.
The only reason these figures were somewhat "disconcerting" was that this period
was characterized by increasing prosperity and developing public health services,
while mortality rates in major European countries showed no such rise. How was
one to explain the American data attributable mainly to degenerative diseases?
There were first of all the standard explanations: industrialism, urban life, personal habits, and pace of life. "This has been the explanation offered by those
identified with the Life Extension movement and by the advocates of periodic
physical examinations generally . . . Their statements have in common the basic
assumption that there has been a real and significant deterioration in our national vitality at the ages under consideration." Dublin, however, urged "great
caution" in accepting this conclusion and argued that there was no mortality increase to explain. [48] He repeated his view that changes in the race composition in
New York State had affected mortality. But he also pointed out that statistics were
largely driven by improved technical procedures in statistical offices, changing
disease categories, and errors of diagnosis and went on to show how these factors
operated for specific diseases. Much of the apparent increase in cerebral hemorrhage and apoplexy was directly related to the radical decline in the use of the
term *paralysis*. The same was true of dropsy as well as conditions formerly classified as "ill-defined organic diseases" that were now being ascribed to "organic
diseases of the heart." He concluded: 'On the main subject at issue I am con-
strained to conclude that there is no evidence to support the contention that the
true mortality rate of the country has increased after age 45."[49] It was perfectly
true, he added, that sharp decreases in mortality in earlier stages of life did not
continue after that age. But it was unreasonable to expect that they would. "Im-
mortality is not one of our characteristics. Persons who do not die from the infec-

tions and accidents at the earlier ages will necessarily come to death at the more advanced ages. This is no cause for apprehension from the health or any other community standpoint." The conclusions were clear. "It is high time, therefore, that the ghost of these degenerative diseases was laid and the program for life conservation at middle life and at the older ages was established upon sound scientific premises. It is not necessary to urge a worthy cause on false assumptions." While it was desirable to call attention to bad habits and to promote annual physical examinations, the program of life conservation should not be dependent on misleading statistics.[50]

As we shall see, Dublin was defending what would eventually become public health orthodoxy in Britain and a minority opinion in the United States. But he remained firm in his convictions. In a study of MLI industrial subscribers published in 1930 but dealing with the situation during the early 1920s, Dublin argued that except for cancer, influenza, automobile accidents, and diabetes, mortality rates were going down for nearly all diseases, including very slightly those of the heart. Although Dublin was disturbed by higher rates for industrial workers, which he ascribed to industrial hazards, death rates had declined and life expectancy had increased even for this group. His outlook remained resolutely optimistic.[51]

The Interwar Years: Extending the Periodic Examination and the Lifespan

After World War I, the campaign for periodic examinations intensified. The *New York Times* continued to publish advertisements by the LEI.[52] Leaders of the organization like Fisk and Fisher continued to publish on the virtues of periodic examinations while speculating just how much longer life might be extended. Representatives of the insurance industry continued to see periodic health examinations as "[o]ne of the most useful disease preventive measures"; many companies were now offering them in one form or another to their policyholders. By 1926, forty-two insurance companies were using the services of the LEI and Lee Frankel could report that among MLI policyholders examined, mortality was 18% lower than expected and the greatest saving was between the ages of 40 and 60. [53] Fisk added a new term to the lexicon of life extension when he argued that chronic and degenerative diseases were *silent sicknesses*; unlike communicable diseases they usually remained undetected for long periods without signs or symptoms. The periodic health examination was "the only resource that we have to make silent sickness speak."[54] And his continuing skepticism about mortality statistics notwithstanding, Louis Dublin also advocated health examinations as an important public health measure that could reduce mortality among the middle-aged.[55]

But the real story of the periodic examination during the interwar period was its extension to new domains. The greatest success of this strategy was the widespread adoption of medical examination of schoolchildren, which became routine in many states. This was viewed both as a model for adult health and as a measure that would protect children from many diseases of later life. But although it was linked conceptually to life extension, examination of schoolchildren had a different institutional focus—the public and private agencies of the child health movement. More directly relevant for adults, the Boston Dispensary in 1918 added a health clinic so that patients who were not acutely ill might have periodic physical examinations. The state of North Carolina also established several clinics for the periodic examination of its citizens. In 1920 Michael Davis, former director of the Boston Dispensary Clinic, established a similar clinic in New York City. That same year, Herman Biggs attempted to get New York State to pass legislation creating health centers, whose functions would include periodic examinations.[56]

In 1922 and again the following year, the AMA's annual convention endorsed the movement for periodic examinations. The ubiquitous Haven Emerson of Columbia University's School of Public Health prepared standardized report forms as well as detailed instructions on how to conduct these examinations.[57] The issue was discussed at many state and local medical meetings. The Conference of State and Provincial Health Authorities of North America endorsed the movement and state health officers were invited to form state committees. In 1923 the AMA's Council on Health and Public Instruction was directed "to initiate and carry out an intensive campaign of education of physicians in the necessity and method of conducting thorough and accurate medical examinations of apparently healthy persons, through the agency of state and county medical societies, hospitals and medical schools." Letters were sent to the leading university medical schools, urging them to emphasize the importance of and provide instruction in conducting such examinations. Letters were also sent to selected hospitals, advising medical staffs to offer local physicians brief courses on how to examine the healthy.[58]

The response to these letters and of the profession more generally was disappointing. The head of the Medical Society of Pennsylvania's Committee on Public Relations complained in 1925 that doctors were slow to adopt examinations and attributed their reluctance to inertia, traditional concern with cure and relief rather than prevention, and the length of time needed to conduct them properly (about a full hour, he estimated). This was unfortunate, he continued, because examinations were a powerful means of binding skeptical patients to doctors. "The claim which most often leads patients to resort to cultists is that physicians do not make adequate examinations. The practitioner who induces his clientele to have annual physical examinations is doing valuable work in preventive medicine, and in addition is attaching his patients to him by the firmest possible bond, that of enlightened self-interest."[59]

Still all was not bleak, he went on. Industrial employers in large numbers had adopted examinations for employees. The army and navy as well were requiring annual examinations of officers. Even the Philadelphia Board of Education was submitting its employees to periodic examinations. Unlike many other doctors, the author was not particularly critical of the LEI, which had "familiarized a large section of the general public with the movement in a most practical way" as well as "providing several thousand general practitioners of medicine with an opportunity to perfect themselves in the technique of making these examinations." But for the most part, he continued, institutional examinations were vigorously opposed. To keep examinations out of the hands of free clinics and dispensaries, the AMA resolved "That the members of the respective societies be requested to make such examinations in the homes or in their offices, free to any persons who, by reason of economic conditions, require such favorable consideration, and that in the performance of the work the same sympathetic confidential relation be maintained between physician and patient or family as has ever characterized the efforts of true physicians."[60]

Doctors, especially in the New York City area, responded in a modest way. The NYAM offered lecture series for doctors on preclinical symptoms of disease and some state and local societies also pushed periodic examinations among their members. They emphasized the advantages of keeping such examinations in the hands of general physicians who, unlike institutions and clinics, knew clients and their families personally. The Judicial Council of the AMA strongly argued "that the family physician is the proper person to perform the health examination." Among their other defects, spokesmen for the Brooklyn Medical Society insisted, LEI examinations consistently found much worse results than did the physician examinations "because they use so much fear appeal in getting their clients. Likewise, not being interested through personal contact with the client their tendency is to give the darkest interpretation to any symptoms found."[61]

The most far-reaching action was carried out by a new organization, the National Health Council. It was widely recognized that fragmented organizations and agencies, both locally and nationally, needed to be brought together to co-ordinate their activities.[62] Various local councils were created (as we shall see in chapter 3) and on the national level, the National Health Council was formed in 1919; this was made up of seventeen major national societies, among them the American Red Cross, the APHA, the LEI, and the PHS. Lee Frankel as president of the APHA was appointed chairman of the council and was authorized to appoint a committee to prepare a constitution and bylaws.[63] One of the new organization's first activities was to announce in 1922 the launch of a national health examination campaign the following year.[64] It was supposed to last only one year under the slogan "Have a Health Examination on Your Birthday" but was continued for a second year. The council produced posters and MLI produced a motion picture called *Working for Dear Life*. Fifty-three prints of this short film were in constant circulation and a set of lanternslides was also prepared. The director of the program, James Tobey, who became known for later work on medical law, echoed a resolution of the APHA by claiming that health examinations and the advice they engendered could add twenty years to the average person's lifespan.[65] Doctors would also benefit by demonstrating "that the modern doctor of medicine practices preventive as well as curative medicine, and that he is more than a dispenser of pills, that he is a purveyor of positive health." While many defects could be found that might be remedied, "The great value of the health examination is, of course, in coping with the organic diseases of middle and later adult life. Periodic human stock-taking should be indulged in at all ages . . . but after thirty-five or forty they are especially useful in bringing to light degenerative diseases in their incipient stages. Such chronic maladies as heart trouble, kidney diseases, cancer, diabetes, apoplexy, and the like seldom give any warning during their beginnings and only an expert can detect them by means of a careful physical scrutiny."[66] Once again it was the danger of so-called organic or chronic diseases that particularly justified periodic examinations.

Periodic examinations were enthusiastically supported by the APHA, which in 1925 undertook a survey of 138 health departments and agencies regarding examination practices. By 1930, a variety of other groups had come to identify with life extension and health examinations. Nurses, especially visiting nurses, it was recognized had a vital role to play in observing damaging conditions and behaviors and providing advice because they spent far more time with patients than other professionals.[67] Nor was the United States the only North American country that was seeking to introduce health examinations. In 1929 the Canadian Department of Health cooperated with the Canadian Medical Association to produce and send out to every physician in Canada a printed examination form. As in the United States, physicians seemed reluctant to use it; "the physician trained in the curative school frequently looks askance at the prospective client who in answer to the question, 'What is your complaint?' is unable to provide a description of an ache or a pain. The public, still unaware of the advantages of the scheme, on the whole stand aloof."[68]

Not surprisingly perhaps the one place where the health examination movement seemed especially successful was New York City. The NYAM and the Medical Society of Brooklyn were particularly active in promoting examinations. In the summer of 1929 the *New York Times* reported that a recently created organization, the Greater New York Committee on Health Examinations, had announced for the late fall "a concerted effort to prolong the lives of New Yorkers through disease prevention and health examinations." Joining the committee in this effort were the New York Department of Health, the Medical Society of the State of New York, county medical societies, and private organizations like the LEI. "Physicians will display posters in their waiting rooms, as well as descriptive booklets on their tables. In all, 50,000 posters will be used in physicians' offices, settlements and schools. A million leaflets will be distributed through schools, colleges and before assemblages of every sort. Prize essay contests are to be conducted in schools, winning essays to be published in leading journals."[69]

The chairman of the Committee, A. J. Rongy, portrayed the periodic examination as "a method through which the old family practitioner, now fading in the light of specialization, may rehabilitate himself and again be the backbone of the community's health."[70] And its secretary, Iago Galdston, characterized it as "the first practical step which organized medicine has undertaken in an effort to save the general practitioner from extinction and change him from an emergency doctor into a health counselor." It was part of the growing trend from "curative to preventive medicine."[71] By April 1930, the committee could announce that the number of health examinations reported by doctors had increased by 25%. And by August, *the New York Times* announced a five-year campaign to promote examinations using films and radio.[72]

Despite all this activity, campaigns for periodic examinations slowed considerably during the 1930s. The public and a majority of doctors, it seemed, showed little interest. By 1942, Charles Winslow was analyzing this failure and that of preventive medicine more generally. Active promotion by medical societies, insurance companies, and voluntary agencies had failed to generate interest, he suggested, because prevention remained "ancillary" to therapeutics in routine medical practice. If less than one person in ten (including infants) in the United States had an annual health examination, it might be because an examination was either cursory and of dubious value "or a fantastically costly procedure with disproportionate results."[73] Although Winslow admitted that this form of prevention had been successfully promoted for children, adults seemed to resist submitting themselves to it. Perhaps, he suggested, rather than aiming at healthy adults, it was necessary to change course and encourage "prompt recourse to the doctor when some impediment to the flow of normal healthy life is apparent to the individual himself. It is when this situation is present, when a man feels that he needs help, when he has been humbled by nature, that he will seek and follow guidance. Counsel on hygiene to those who feel wholly adequate has little effect; but when something hurts, or when activity and efficiency are consciously limited, then, with some degree of reluctance, varying with the individual, we are willing to seek advice."[74]

Conclusion

The interwar campaign for physical examinations while not a great success was hardly an insignificant failure. It contributed greatly to early concern with chronic disease because it promised to uncover hidden and terrifying conditions before they became untreatable and possibly fatal. It brought together individuals and groups pursuing different and sometimes conflicting aims. It promoted a new form of prevention that differed markedly from traditional public health activities and brought them into contact with traditional medical practice. It was an important component in the development of mass campaigns to influence public behavior. It permanently marked health policy in the decades to come and would resurface in various guises. The idea of periodic physical examinations would become an essential element of several wider movements: efforts to expand the scope of public health activities; New York City's massive campaign to rationalize its hospital and welfare institutions; and, not least for the purposes of this book, the campaign to confront chronic disease.

CHAPTER TWO

Expanding Public Health

In 1920 C.-E.A. Winslow, professor of public health at Yale University, presented a long address to the American Association for the Advancement of Science in which he made the case for a "New Public Health" with a vastly enlarged role. Such expansion was already underway, he argued. Public health had moved from environmental sanitation to control of communicable diseases and, most recently, education for individual hygiene. While there existed general principles of hygiene that applied to everyone, there were also specific measures that applied to each individual. This required putting an end to the artificial division between prevention and therapy because diagnosis was necessary to determine the individual's physical and mental state and such diagnosis frequently demanded therapeutic intervention. The conflation of prevention and therapy was already taking place in child health, tuberculosis, and venereal disease programs. "If it is sound economy to provide for the early diagnosis and sanatorium treatment of tuberculosis, it is just as 'sound' to provide for the early diagnosis and surgical treatment of cancer. The two diseases are equally dangerous, and equally burdensome to the community; they are equally preventable, if the right educational and clinical procedures are organized for their control."[1]

When Winslow wrote these words, the US Public Health Service (PHS) was a small agency doing routine sanitation and inspection work. But it was on the cusp of change. The New York City Department of Public Health was setting new standards in expanding the scope of the field. The creation of schools and academic departments of public health at major universities during the second decade of the century brought a new degree of expertise and research competence to the field.[2] The pages of its journal demonstrate that from 1920 on the members of the American Public Health Association (APHA) devoted much time and effort to the task of defining, standardizing, and evaluating public health work. The move to take on chronic disease was part of this larger endeavor.

Public Health during the Interwar Years

A leading figure in this movement was Herman Biggs. After he left New York City in 1913 to become director of public health for New York State, Biggs remained an influential presence until his death in 1923. His views on the need to expand the scope of public health grew more ambitious. In 1920 he was responsible for the introduction of a bill in the New York State legislature to allow county health units to establish "health centers."[3] The idea was to permit local authorities to bring together the disparate health services in every district, from public health nursing to hospital surgical treatment. Included in these services was an annual health examination to detect physical defects and diseases and to suggest methods of correction. Local representatives of public health agencies were to play a supervisory role. The state government would provide grants to these local centers, subject to inspection by state health departments. The result would have been an integrated system of preventive and curative medicine under state auspices.[4] Instead of providing health insurance coverage as others were advocating, the "health center" bill would provide services free to the poor based on a means test, while those who were able paid for care and treatment. Milton Terris saw the 1920 bill as Biggs's rejection of health insurance in favor of an extension of state medicine with salaried physicians.[5]

Biggs himself framed the bill as follows: "It may be considered a practical form of governmental aid to the practice of medicine applied thus far almost wholly to the infectious diseases. The wide development of public health laboratories in recent years is a notable incident of this tendency."[6] Although the bill received support from a number of leading public health figures, including Charles V. Chapin, superintendent of health in Providence, Rhode Island, as well as from interest groups like the State Federation of Labor, it also provoked considerable opposition from the medical profession and was defeated. The bill said not a word about chronic disease or disability.[7] In his last major address delivered in 1923, Biggs devoted attention to cardiovascular disease, cancer, and diabetes but in the context of a vast program of health promotion centering on regular physical examinations.[8] Nonetheless, the campaign to expand public health by coordinating healthcare and prevention activities would serve as a framework for subsequent concern with chronic disease.

Biggs was in a minority within the public health movement, but he was not an isolated figure. The developing world of academic public health produced a variety of spokesmen defending such views, often in the most extravagant terms. In 1915 W. T. Sedgwick, professor of biology and public health at MIT, wrote that for most of the nineteenth century public health could boast of few preventive successes aside from smallpox vaccination. It had more recently added preventive sanitation, like the purification of water, sewage, and milk; control of mosquitoes to prevent malaria and yellow fever; and some improved housing.

> But we have not yet even begun to demand that study and care of the individual which is still more fundamental. We have as yet paid little or no attention to overeating, overworking, over-drinking, deficient exercise, deficient sleeping, family hygiene, and the hygiene of special organs, such as eyes, ears, bowels, nose and feet, all of which I propose to group together under the term preventive hygiene . . . Rightly studied, preventive hygiene will include personal, domestic, family and social hygiene. It will deal with celibacy and marriage, with housekeeping, with the high cost of living, with food economy, with domestic service, with child hygiene, and with the right conduct of mature and elderly life, as well as with personal hygiene. It will in the future play, perhaps, the principal part in solving many of the problems of American life, health, prosperity and happiness.[9]

Such grandiose claims became even more common following World War I. In 1921 Frederick Gay, a bacteriologist working in California, told the Pacific Division of the American Association for the Advancement of Science that it was the function of public health to spy out and remedy the "ills that flesh is heir to," and to "deal with the individual and collective problems of disease, ignorance, vice, crime and poverty. It is evident we have here the whole tissue of human altruism, and have far outstripped the meaning of public health in common speech."[10] Others advocated more targeted objectives. Eugene C. Howe of Wellesley College and MIT focused on the apparent increase in death rates among middle-aged men and emphasized that boards of health had "turned resolutely" to promoting the practices of personal hygiene. He suggested that a key element in this strategy should be hygienic training in schools, supervised by teachers of physical education. In 1919 Lee Frankel of Metropolitan Life Insurance Company (MLI) emphasized the need for a concerted, unified, comprehensive federal health program that was centralized and coordinated by a "progressive" federal health agency with adequate staff and financial resources.[11] In that same year, Frederick Hoffmann, of the Prudential Life Insurance Company and a noted statistician and controversial public figure, wrote: "Under such a conception the modern health department would assume the functions of a general health administration concerned with all the matters which affect the health and physical welfare of the people."[12]

Some of these calls to expand public health were direct responses to what appeared to be the rise of chronic diseases. For instance, Aron Arkin, professor of pathology and bacteriology at West Virginia University, wrote about the apparent rise in mortality from cancer and cardiac, vascular, and renal disease and called for an integrated health system: fully equipped health centers in every community and available to the family physician for the diagnosis and treatment of disease, and periodic physical examinations; group medical clinics in cities for more expert diagnosis and treatment; development of federal and state programs in disease prevention; and the promotion of scientific research and investigation. Modern medicine, he concluded, "is becoming more and more a social service concerned with the prevention of disease, prolongation and betterment of life, improvement of physical health and efficiency through organized community efforts."[13] The APHA in 1922 adopted a resolution authored by Herman Biggs, Lee Frankel, and Haven Emerson to the effect that the goal of the association should be to add twenty years to average life expectancy in the United States within the next fifty years.

> It is the opinion of the American Public Health Association that the maximum life expectation is far from having been attained, even with no further additions to our knowledge of the cause and means of prevention of disease. By adding to scientific control of communicable diseases and the protection of infancy, the avoidance of disorders of nutrition and the degenerative diseases of middle age we may well promise the attainment in the next fifty years of a span of healthy life beyond the scriptural ideal of three score years and ten.[14]

After Biggs's death, the leader of the expansionist movement in public health was unquestionably C.-E.A. Winslow, who in 1915 became professor of public health at Yale. Educated at MIT, where he had worked with Sedgwick, Winslow reflected the changing orientation of public health. Educated in biology and for many years editor of the *Journal of Bacteriology*, he turned early in his career to health education. From 1910 to 1915 he was curator of public health at the American Museum of Natural History, where he organized health exhibits. In 1914–15 he worked briefly under Biggs (about whom he later wrote a book) as director of the Bureau of Public Health Education in the New York State Health Department before founding the Yale Department of Public Health. Among other activities early in his career, he wrote a book for children on the body and how to care for it that was published in numerous editions.[15] If he emphasized education for individual hygiene, Winslow was nonetheless aware of the need for social and environmental action though he was vague about mechanisms. For the "New Public Health" to be effective, it was necessary to provide wider access to medical care. Something had to be done to mitigate poverty, closely correlated to disease, so that people could live under healthy conditions. He was equally vague about what to do about "unfit" individuals who could not be helped by such measures but hinted at some sort of eugenic solution. His definition of public health was not modest.

> Public health is the science and the art of preventing disease, prolonging life, and promoting physical health and efficiency through organized community efforts for the sanitation of the environment, the control of community infections, the education of the individual in principles of personal hygiene, the organization of medical and nursing service for the early diagnosis and preventive treatment of disease, and the development of the social machinery which will ensure to every individual in the community a standard of living adequate for the maintenance of health.[16]

Nor was he modest about the potential benefits. "The science and the art of public health have progressed to a point where they can render to the public a service to be measured in the saving of hundreds of thousands of lives in this country every year."[17]

Several years later, Winslow published a short history of public health, which in his account had gone through several stages. There were the "dark ages" before modern public health, followed by the sanitary awakening and the bacteriological era. Developments culminated in the "New Public Health," which according to Winslow had its origins in the tuberculosis movement. Once it was established that open-air therapy and proper regimen could cure many cases, sanatorium treatment developed, along with dispensaries and homecare. At the same time, the discovery by tuberculosis pioneers that the "education of the individual in the practices of personal hygiene" was the essence of preventive medicine "has proved almost as far-reaching in its results as the discovery of the germ theory of disease thirty years before."[18] The same insight recurred in other domains: the infant care movement; the medical supervision of schoolchildren; syphilis prevention campaigns; and the mental hygiene movement. In each case, "the fight must be won, not by the construction of public works, but by the conduct of the individual life."[19] The public health nurse was the chief agent for carrying the hygienic message into the home while physicians fostered prevention through medical examinations of well persons in order to diagnose diseases in early or incipient stages. Together they provided the basis of the modern public health campaign.[20] Recent efforts to face the degenerative diseases of adult life "aim to bring within the range of preventive medicine diseases which the health worker of 1900 would have unhesitatingly classed as non-controllable. The possible field of public health is thus extended so that it is almost co-extensive with the range of physical liability."[21] Any program of systematic medical examination of the population over 45 years of age required new methods of financing healthcare since most individuals would not pay to visit physicians unless they were visibly ill. In discussing health insurance, Winslow seemed to come down on the side of its critics, who believed that it underemphasized prevention; he favored instead community health centers like those Biggs had tried to establish in New York State. Notably missing from his 1923 book is the role he had earlier assigned public health of dealing with poverty, work conditions, and poor housing. Instead, Winslow invoked declining death rates to illustrate the power of public health to transform human life. Triumphalism, rather than anxiety, fueled the ambitions of the "New Public Health."

In his presidential address of 1926 before the APHA, Winslow began in the same triumphalist tones. The results of public health during the last fifty years "constitute without exaggeration one of the most startling and revolutionary events in the whole history of the human race." Nonetheless, a new problem had emerged. During the previous half-century, communicable and environmental diseases "have been so substantially reduced that the problems of the future are heart disease, the acute respiratory diseases and cancer. We face a new situation and we must adopt new methods if we are to meet it with any measure of success."[22] He went on for the first time to discuss these newly problematic illnesses at length, starting with organic heart disease, moving on to cancer, and then to mental disease. In all cases, the essential issue was the education of the individual not in abstract principles but through "the preventive application of the resources of medical science to the individual case."[23] If his solution had changed little from earlier talks (excepting the elimination of the socioeconomic concerns in his 1920 paper), Winslow had clearly thought through a number of key points.

He rejected the easy distinction between prevention as a task of the state and treatment as the responsibility of the private physician because in the case of degenerative disease cure of an early disease amounted to prevention of more serious damage and the "promotion of physical and mental health and efficiency."[24]

While the resources of sanitary engineering and bacteriology had been made available to the entire community, the major problem of health conservation had become the effective application of the resources of medical science to individuals. "It is on such an application that the control of heart disease and cancer depends."[25] And making medical care truly preventive required a "fundamental change in the system of compensation for medical service." While he remained open about the form that such change should take, a recent visit to England had convinced him that the worst problems of the health insurance program had been resolved. But he still saw the health center movement as the most promising direction for the United States. Mental hygiene, heart, and cancer clinics would join existing pre-natal clinics, school medical services, and clinics for tuberculosis and venereal disease. Some form of organized medical services, he opined, is "coming as surely as the sun will rise tomorrow." This required cooperation with the medical profession and Winslow urged the APHA to undertake a systematic study of the question and to request cooperation from the American Medical Association (AMA).[26] Winslow's abandonment of issues of poverty and social change that had so animated Herman Biggs[27] and that he himself had espoused more mildly only a few years before was characteristic of much public health thinking of this decade. The onus now fell on "the intelligent self-control of the individual."[28]

Some influential individuals agreed with Winslow. In 1926 the surgeon-general, Hugh Cummings, quoted with approval the finding of a survey of 1921 on municipal health department practices to the effect that the municipal department of health should provide leadership for community programs of hospital development "because in the conduct of its own work it needs hospital facilities of various kinds," and because "it is the one agency supported by all the people to which the people may look, or should be able to look, for guidance." The primary function of health departments should be "the correlation, coordination, and cooperation of all the actual and potential agencies of public health."[29] A year later, Cummings advanced a more general vision. Pointing to the growing importance of "degenerative diseases of advancing life," he rejected the separation of prevention from cure as an error. "We are coming more and more to recognize the unity of medicine and more and more to see that if we are to achieve the desired results, all departments of medicine must cooperate to the one end, which is the prevention of disease and the promotion of health."[30] Although this became the conventional wisdom of the public health establishment during the 1930s, such rhetoric left open the question of who did what and the degree to which curative medicine would be coordinated, organized, or carried out by public health agencies.

The Internal Opposition

This expansive vision of public health did not go unchallenged within public health circles. Opponents of the move into treatment services argued that public health would lose sight of its basic goals by expanding its scope, "or be rendered inoperable if it chose to confront organized medicine." It was suggested that local health officers, who "served at the sufferance of local medical societies," were already "fearful of accepting any variance in a traditional interpretation of public health."[31] At a symposium on the future of public health in 1928, Matthias Nicoll Jr., commissioner of the New York State Department of Health, stated plainly and clearly: "Curative medicine, in the commonly accepted meaning of the term, should not be included among official public health activities, except in emergencies and when, in the absence of proper medical care, there arises a menace to the health of the community."[32] Responding to increasing public pressure to do something about cancer, Eugene R. Kelley, state commissioner of health in Massachusetts, distinguished in 1924 between "public health problems and what are problems of practical public health administration." Cancer was undoubtedly a huge public health problem. But no problem, no matter how grave, could become an administrative responsibility until it was possible to deal with it in an effective and practical way "that will give assurance or at least great promise of good return for the money expended, keeping constantly in mind how far that same amount of money might go towards achieving permanent results in other lines of health administration."[33] And this was manifestly not the case in Kelley's view insofar as cancer was concerned. The steps that needed to be taken— research into causes and adequate hospitalization—did not fall under the jurisdiction of public health administration. The state agency was capable of pursuing statistical studies of prevalence and cure and encouraging periodic examinations and use of state diagnostic facilities. The most critical task of health departments, he suggested, was to mount effective educational campaigns in order to combat fatalistic attitudes toward illness.

The most influential opponent of the new, expanded public health was almost certainly Haven Emerson, briefly commissioner of health in New York City, longtime professor of public health at Columbia University, and a pervasive presence in American public health, arguably as influential in his way as Winslow. Emerson's disagreement with Winslow was explicitly ideological; his critique focused on what was being perceived as the problem of "adult" health, another way of talking about chronic disease. He intoned: "The resources for future benefit to adult health will be largely those which depend on personal initiative and acceptance of responsibility for their cost, rather than through expansion of public agencies, official or volunteer." He then added: "A higher standard of living, a simpler philosophy of life, and more of the quality of social justice will be leading general factors." He suggested that improvements in adult health should be based on the same factors that had led to health improvement in the past: hygienic education; protection of child health in order to prevent diseases of later life; and control of communicable diseases, of the environment, and of the conditions of work. There was no evidence that the health of people over 45 had deteriorated as had been widely believed a decade before. "A more mature, longer and less biased study shows that every age shares to some extent in our advancing safety of life." Nonetheless, he argued, sounding like an early version of Thomas McKeown, "there is little honest record of cause and effect which can be conjured into a safe proof that any particular effort of organized health work has materially improved adult health."[34] Any health services that had been introduced were of far too recent origin to have made a significant difference.

Future progress would depend on a "rational philosophy of life, a physiological concept of existence rather than by the introduction of new functions into official or volunteer health agencies." This strong belief in individual responsibility for health was at the core of Emerson's commitments and went beyond responsible behavior to include individual self-financing of prevention as well as healthcare.

> I question the soundness of any policy of official or volunteer service, whether for adults or children, which for the sake of quick and easily proved results, relieves the beneficiary or the guardian, of the primary duty of seeking and paying for the service. Industrial hygiene provided by the employer is less valuable than when organized as self-protection by the worker. School health service will continue to be a makeshift until parents themselves provide the equivalent of it for their own children at their own expense.[35]

He then added a final thought. "In concluding, may I be permitted to suggest that sociology, economics, psychology and international law are destined to make as much, if not greater, contributions to adult health through the established channels of popular education, than bacteriology and sanitary science have in the past."[36] Years later, the commissioner of health of Baltimore emphasized the limits of public health by invoking the necessary separation and cooperation between public health and medical work. No health department, he insisted, could do its job without the active support of the medical profession.

Such work as the eradication of diphtheria, the reduction of the venereal diseases, and the adequate care of expectant mothers and of young children must rest largely in the hands of the practising physicians. A health department can do much to teach the public to seek such keep-well services from the medical profession. It should never hope to have a staff or clinic service large enough to carry on the needed volume of work itself. The practising physicians are the outposts on the firing line in the modern battle for better health and they should be considered by all as veritable public health agents.[37]

Despite such differences, there was much that united virtually all public health figures. Few would have taken issue with Winslow's statement of 1929: "The object of our whole program is, I take it, to change the conduct of individual men, women and children. We want them to manage their bodies well."[38] No one seemed to doubt the importance of workplace conditions, although it was less clear at whose feet blame for inadequacies and responsibility for change should be laid. There was equally consensus about the central role of physical examinations for adults. Any differences of opinion had to do with how these should be organized. Should everything except public advocacy for examinations be left to the medical profession; should public authorities limit themselves to diagnosis and refer the sick to physicians; or should health departments set up comprehensive public programs similar to those for child health, tuberculosis, or venereal disease? Some thought that hospitals should play a central role through outpatient programs.[39] Many of those who insisted on the central role of private-practice physicians were themselves MDs like Emerson, who held strong ideological commitments opposing state encroachment on healthcare. For others, physician examinations were a way of attenuating increasing tension between doctors and public health workers. As we saw in chapter 1, the APHA in 1922 first endorsed the principle of annual physical examinations by physicians. One public health officer admitted three years later that public health advocacy of periodic examinations had done much to alleviate doctors' suspicions about public health officials. He cited his own situation where his promotion of examinations "has convinced the physicians of his district that there is no plan afoot for making medicine a public utility, with the result that this health officer enjoys the full confidence of the medical profession and receives hearty cooperation in every public health activity that he undertakes."[40] Several years later, the Detroit Bureau of Public Health tried to mollify the city's increasingly hostile medical profession by introducing a campaign of prevention centered on convincing the populace to visit doctors for a variety of preventive services.[41]

Despite or perhaps because of the disagreements surrounding the therapeutic role of public health institutions, periodic physical examinations became a central pillar in the program to expand public health. Here was something everyone could agree on, so long as one remained flexible about the forms of provision. Physical examinations were less closely associated with chronic or degenerative diseases in this public health context than in the case of the life extension movement because they could be applied to a wide range of patients and conditions. But chronic diseases were nonetheless becoming prominent as public health issues. The chief engine for such concern was growing popular anxiety and fear surrounding cancer.

From Cancer to Chronic Disease: The Case of Massachusetts

The relationship between cancer and chronic disease as objects of concern was complex. Many of those associated with the chronic disease movement considered cancer to be a critical part of their mandate. In some ways, it was *the* emblematic chronic disease, having evolved from an incurable condition to one that could now be treated. But those wishing to advance cancer research and treatment, a somewhat larger and generally more affluent group, did not necessarily think of cancer as part of some larger entity or even as a real chronic disease. (Patients either were saved or died.) The groups around the two diseases can best be conceptualized as two distinct and separate social movements that occasionally intersected and infrequently provided mutual support for one another. The two movements, moreover, were not evenly matched. Cancer provoked individual terror and anxiety on a sizable scale. Chronic disease was an abstract second-degree category that meant little to most sick individuals, if it was not actively stigmatized as a problem of poverty or old age. It had salience mainly for individuals and groups interested in large-scale organization and management or medical researchers attracted to a variety of diseases that were poorly understood and even more poorly funded.

Much has been written about the development of the cancer movement in the United States. It provides an example of a general trend in twentieth-century American medicine—"the rise of nonprofit health interest groups focused around a single disease such as tuberculosis, polio, cancer, diabetes, and AIDS and the alliance between these interest groups and Congressional leaders to promote research, treatment and other favors for those afflicted."[42] It is unlikely that this disease was more "dreaded" than it had been in the nineteenth century. It was certainly more visible, however. Statistics indicating that cancer mortality was rising continued to be hotly contested by statisticians, but controversy did not make the unease they provoked less intense. What changed most radically were the attitudes surrounding cancer, a disease previously thought to be incurable and in fact shameful.[43] By the early decades of the twentieth century, many believed that cancer could sometimes be cured through early detection, surgery, and especially, radiation therapy. And even in the many cases that remained incurable, palliation and extending life to the maximum were thought to be more appropriate social responses than allowing destitute cancer patients to die miserably and painfully in poorhouses. Finally, it was believed that continued scientific research would expand the scope of curability and possibly solve the cancer problem. And so organizations like the American Society for the Control of Cancer (ASCC founded in 1913 and eventually to become the American Cancer Society) were able to mobilize large numbers of activists—politicians, philanthropists, and medical institutions like the American College of Surgeons (ACS), which in 1931 produced guidelines on cancer-care wards in hospitals—to create institutions whose mission it was to find and treat cancer patients.

In several American cities, voluntary and public social agencies began coordinating their activities in the early twentieth century. The Boston Council of Social Agencies was founded in 1920 for this purpose and within a year had 176 member agencies.[44] In 1926 the council decided to study the chronic disease problem in the Boston area.[45] The report (to be discussed in greater detail in chapter 3) found considerable incidence of chronic disease and disability among the city's indigent poor and emphasized the need for more numerous and appropriate institutions to deal with this population. There was some response to the report but the problem of chronic disease among the indigent was almost immediately overshadowed by intense public concern with cancer. It was cancer that soon drew the Massachusetts Department of Public Health into the discussion surrounding chronic disease.

As elsewhere, cancer in Massachusetts provoked profound apprehension. From 1915 on, petitions were presented annually to the legislature calling on the Massachusetts Department of Public Health to organize medical care for cancer patients. Calls for state action on cancer took place within the larger context of debate about the role of curative care in public health activities. In 1920 the public health department took over responsibility for four tuberculosis sanitaria and a leprosy hospital. But the director of the department, Eugene Kelley, was wary of such undertakings and opposed the creation of cancer facilities, viewed as the province of private physicians and hospitals; he preferred that the department provide advice, diagnostic services, and information. In 1925 the Massachusetts legislature established a joint committee made up of representatives of the departments of public welfare and public health to examine services for cancer patients and look into the possibility of creating a state cancer hospital. Initially, most of the pressure in favor of such a hospital aimed at providing care for terminal cases. It was argued that although general hospitals were finally treating at least some cancer patients considered curable, there were no facilities for those deemed incurable. This situation gradually shifted as demand focused increasingly on the provision of therapeutic services.[46]

As a result of pressure from the legislature, a young statistician, Herbert Lombard, was brought into the department to do studies on the epidemiology of cancer. Legislators broadened the reform mandate from facilities for inoperable cases to include facilities for diagnosis and therapy. The joint committee produced two contradictory recommendations. Representatives of the department of public welfare advocated the construction of a new 340-bed cancer hospital. Representatives of the department of public health run by Kelly rejected state involvement in medical practice or public care for cancer patients. Needs of indigent people, they argued, should be met by the department of public welfare. Despite this opposition, the state legislature, in 1926, directed the department of public health to set up a network of cancer clinics throughout the state and authorized the creation of Pondville Hospital for cancer patients needing longer-term care; this institution, opened in June 1926 with 90 beds, soon expanded to 115 beds. The role of the department of public health was justified on the grounds that making early therapy more widely available was a preventive measure against late-stage incurable cancers.[47]

Kelley, still state commissioner of health, was not happy about these developments. As we saw above, he recognized cancer to be a public health problem but not one of public health administration. He sat on a special commission on cancer appointed in 1925 by the APHA that made the modest suggestion that state health authorities should make pathological services available without cost to doctors and patients, and should encourage periodic medical examinations to catch the disease at an early stage.[48] Kelley's successor as commissioner of health, George Bigelow, took a very different approach. Two years later, another APHA commission on cancer was struck. It included Bigelow, as well as the head of the

ASCC, George Soper, and proposed a much more ambitious program to the association. Municipal and state health authorities, they argued, should be actively involved in providing facilities and resources to diagnose and care for cancer patients, curing those who could be cured and palliating the rest.[49]

Back in Massachusetts, Bigelow and Lombard adapted in yet another way to the new situation. They linked cancer to the "chronic disease" discussion that was taking place around the welfare issue. During debate about the cancer hospital within the joint interdepartmental commission, public health officials had argued against the creation of a new hospital on the grounds that "if the state offered diagnosis and treatment to cancer patients, these services were equally applicable to other chronic diseases such as diabetes, nephritis, and heart disease."[50] Bigelow and Lombard seem to have taken this rhetoric to heart because once the principle of a cancer hospital and clinics was accepted, Lombard expanded the scope of his epidemiological studies to include chronic diseases more generally. In 1929 Lombard published his first report on chronic disease in Massachusetts. In 1930 both Lombard and Bigelow presented papers on chronic disease at the annual meeting of the National Conference of Social Work (itself an indication of the growing intersection between public health and welfare institutions). Here Lombard argued that the "outstanding public health problem of the present day is the control of chronic diseases of the middle aged."[51] Bigelow had a similar message about cancer, although he could not resist a dig at the lack of interest in other diseases. "I cannot help contrasting the magnificent response to this program with the meager response to our venereal disease needs. Today we spend over a quarter of a million annually for cancer, and less than a fifth as much for gonorrhea and syphilis." Scientific knowledge of both diseases would suggest quite the opposite emphasis, he suggested. He also announced the creation in his department of a Division of Adult Hygiene dealing with other degenerative diseases and promoting interest in health examinations. After the cancer program, he explained, he was attempting "to see ahead into other degenerative disease fields."[52]

Bigelow was not operating in a vacuum. Governor Allen recommended to the Massachusetts legislature an intensive study of problems of aged sufferers from chronic disease, with "the particular purpose of determining how far the insecurity of old age may be lessened by early diagnosis and treatment of cases of chronic disease." The goal was to come up with more adequate facilities. The professor of orthopedic surgery at Harvard responded to the governor with pleas on behalf of sufferers of rheumatic and arthritic illness and noted that others had lobbied on behalf of other chronic diseases.[53]

The results of the Massachusetts survey were published in book form in 1933 and played a major role in turning chronic disease into a genuine public health problem. The volume was notable in several respects. While virtually all previous surveys of chronic disease had targeted welfare populations (as we shall see in the following chapter), this was the first one that looked at the population as a whole. And results, according to the authors, were unambiguous. They began their first chapter with these striking lines: "The problem of chronic disease will not be downed . . . So this thing called 'civilization' is working poor old Nature overtime. Increasingly great numbers of people are ill, crippled and dying from chronic disease, and the problem thus created will not be downed."[54] Wading into the debate about whether the apparent rise in incidence of chronic disease was real or an artifact of changing age profiles and diagnostic techniques and categories, the authors argued that the rise was real and "it would be the height of folly to refuse to face the problem." Their own survey data had found that about 12% of the state population suffered from some form of chronic disease; 45% of these sufferers were partially disabled and 5% were totally disabled. Rates were especially high among those who were over 40 and poor, as well as in families cared for by public welfare departments. Consequently, chronic disease was a very costly problem. The data, moreover, underestimated incidence since it was based on responses to survey questions; not everyone recognized their sickness and some deliberately falsified responses. The authors estimated that in the entire state between 475,000 and 560,000 individuals suffered from chronic disease at any one time.[55]

More interesting than the data, which would soon be overshadowed by the National Health Survey (NHS), was the argument that public health had to take charge of chronic disease. This position was based almost entirely on the Massachusetts experience with cancer advocacy. In the first sentence of the preface, the authors wrote: "The question of the scope of public health has long been a subject of dispute. Each new addition to the field has brought with it contention from those intimately interested." Modifying Winslow's iconic periodization of public health eras, they suggested that public health began with sanitation, was extended to bacteriology, then personal hygiene, and "in the last few years has gone into a modified form of state medicine." (They were referring to public provision of healthcare for maternal-infant health, tuberculosis, venereal disease, and mental illness.) And the reason for this was simple. "The people demand more and better service from government."[56] Physicians felt that government was encroaching on their territory and health officials believed that cancer care was out-

side the scope of public health, unlike treatment for venereal disease that eliminated spreaders of disease. But such reservations could not stand up to public and
political opinion. The legislature had spoken and soon after the APHA had taken
the stand that cancer was a legitimate public health activity. State and city programs followed. The fact is, Bigelow and Lombard argued, "the outstanding sickness and health problem of the present day is the control of chronic diseases of
the middle-aged." They added: "There is hardly a family in Massachusetts without immediate experience with cancer, heart disease or rheumatism."[57] It was
thus not surprising that an emotional public response was being generated that
public health institutions ignored at their peril. The cancer program demonstrated that "When aroused the legislature will demand service for a chronic
disease." There was no reason "why similar services under such auspices should
not be developed for other of the important chronic diseases . . . There is reason
to suppose that if someone else does not take the initiative the people through
their legislature will again do so, to the great shame of the medical and public
health professions." Public health officials in the United States and Canada had
little interest in adult hygiene and chronic disease, often due to fear of antagonizing the medical profession. "But they seemed to have no fear of antagonizing
the public on whose good will their livelihood depends."[58] It was imperative to
develop close cooperation between public health and the medical profession in
dealing with chronic disease. The authors quoted W. S. Rankin, now at the Duke
Endowment, and set themselves firmly in the tradition of Rankin and Winslow:
"the forces of medicine and public health cannot be separated along lines of cure
and prevention, as neither medicine nor public health can afford to renounce its
interest in either treatment or anticipated treatment. Cure and prevention merge
by imperceptible gradations as physiology becomes transformed into pathology,
as the new leaf of spring fades into the seared yellow leaf of autumn."[59]

It is thus somewhat ironical that the cancer treatment program was not followed by other chronic disease initiatives in Massachusetts, probably because the
state experienced little of the public pressure to confront other chronic diseases
that Bigelow and Lombard had anticipated. Nothing like the broad program established in New York City was created in Massachusetts before World War II.
Nor were significantly more facilities for the chronically ill established despite
a report in 1934 suggesting that little had changed and calling for increased resources and facilities. This last report (ironically, it was written by Haven Emerson) concluded: "Boston lacks sufficient care for chronic disease."[60] When all was

said and done, cancer in Massachusetts (and elsewhere) occupied a special place in the public consciousness and had unique power to mobilize politicians and ordinary citizens. Nonetheless, in the context of the Depression and the New Deal, the chronic disease problem would advance separately but in step with the cancer problem.

Public Health and Biomedical Research

The PHS did not just promote public health practices; it was also the major federal body responsible for medical research. Much of this research concerned infectious diseases, or diseases thought to be infectious like pellagra. But as the mechanisms of infectious diseases became better understood, chronic diseases became the beachhead of advanced research. Here too, cancer received most of the attention. Initially, efforts on the national legislative front in the late 1920s for greater government involvement in cancer research were not very coherent (offering, for instance, financial rewards for a cancer cure) and were largely unsuccessful. There existed within the PHS an existing research sector, made up of a Division of Scientific Research and a Hygiene Laboratory where both applied and fundamental research took place on a small scale. New state needs had regularly expanded the role of the laboratory; a congressional act of 1912 extended its mandate to include pollution of lakes and streams. By the 1920s, the laboratory was doing increased research on diseases like cancer. But private philanthropic foundations and disease-specific associations remained the primary source of funding for such research.

Nonetheless, growing popular faith in science produced considerable pressure for change. A campaign begun in the mid-1920s by scientists (notably chemists associated with the Chemical Foundation) and several members of Congress to increase funding for research led to years of negotiation and political maneuvering that produced the Ransdell Bill of 1930; this transformed the Hygiene Laboratory into the National Institute of Health (NIH), able to accept private donations. The Parker Bill of the same year reformed administrative regulations that allowed the NIH much greater organizational flexibility. Despite initial optimism, the Depression and a conservative PHS leadership opposed to rapid expansion combined to restrain the research activities of the NIH.[61] Nonetheless, some important principles were adopted although not immediately implemented. The importance of fundamental research was affirmed and a decision reached to establish a new Division of Physiology and Physics. Research on noninfectious diseases was increased. The cancer program set up in 1922 was substantially expanded. In 1932 a section for the study of heart disease was established. The NIH had by then laid out a program of increased emphasis on chronic diseases.[62]

The NIH created an institutional focus for demands to increase state intervention in medical research and expand into areas like chronic diseases. For although the problems of infectious disease were far from being resolved, there existed sufficient knowledge to allow for prevention and targeted research. In contrast, the uncharted frontier of medical science now consisted of the "functional," "degenerative," or "organic" diseases whose complex causality was poorly understood. Solving their mysteries seemed to require biological research that was more sophisticated and expensive than anything undertaken thus far. This served the interests of both scientists and PHS administrators seeking expansion and larger budgets. The combination of public fear of cancer, belief in the promise of medical research, scientific interest in the puzzle of chronic diseases, and administrative need to justify the expansion of research activities produced a new emphasis on chronic diseases. In 1934 the President's Science Advisory Board created a subcommittee packed with members and friends of the PHS that called for an increase in the annual funding of the PHS of $2.5 million and for increased research into a number of diseases, including several considered chronic.[63] Nor was such activity confined to the federal government. As we shall see, cities and private philanthropists also began to invest in research centering on chronic diseases.

The effort to expand public health beyond infectious diseases and prevention would intensify during the 1930s as a result of efforts to reform the American healthcare system. The public health leadership appointed by the New Deal adopted the view that expanding research facilities and activities was an integral aspect of the overall expansion of state-supported healthcare and that such research should place increased emphasis on little understood chronic diseases. And unlike other aspects of the expansionist public health program, investment in research threatened few established interests and could be embraced by everyone, including the medical profession. Few other healthcare issues inspired so much agreement.

Conclusion

Developments in Massachusetts or motions by the APHA about confronting cancer or even the expansion of research did not resolve the issue of the boundaries of public health. Opposition to expanding public health activities remained vigorous within the field and without. In 1933 the Committee on Administrative Practice of the APHA (chaired by Winslow) produced a statement that had been prepared by its Subcommittee on Essentials of Health Organization (chaired by Haven Emerson but including Winslow). Titled rather awkwardly "An Official Declaration of Attitude of the American Public Health Association on Desirable Standard Minimum Functions and Suitable Organization of Health Activities," this document listed two primary goals for local public health agencies: (1) the control of communicable diseases and (2) the promotion of child health. Not a word was mentioned regarding chronic illnesses or the involvement of public health agencies in curative medicine. A mission statement of 1940 was similarly restrictive in its language.[64]

Nonetheless, beneath the surface much was going on within both the APHA and the PHS. The top leadership of the PHS in particular was in a position to actively pursue more expansive programs. Surgeon-General Thomas Parran in 1939 bluntly stated: "There was a time when public health activities were limited to the control of epidemic diseases and basic measures to improve the sanitation of the environment. Health activities of today, however, are concerned with the prevention, alleviation, and cure of all of the major causes of disablement and death. Prompt restoration of the patient to health is equally important with the prevention of disease."[65] Such statements represented aspirations rather than re-alities, but they reflected real changes that were taking place under the The New Deal. Before analyzing this shift, however, there remains one final domain to be explored.

CHAPTER THREE

Almshouses, Hospitals, and the Sick Poor

Hospitals in the nineteenth century were charitable institutions for the poor and closely associated with other welfare agencies. By the twentieth century, they were becoming highly medicalized and serving the entire population; but they nonetheless frequently remained administratively and practically connected to the welfare sector. For many in this welfare-hospital domain, chronic care was becoming a central issue. Nearly everyone agreed that the new "modern" hospital was not equipped to deal well with the very poor who did not suffer from acute diseases; but few alternative institutions were available to replace them. Out of this apparent vacuum there emerged a movement to create an appropriate and rational system of care for indigent, chronically ill people. Among those concerned with the issue were doctors and nurses working with these populations and seeking to improve their care, as well as hospital physicians and administrators who, in contrast, viewed "chronics" as a nuisance to be eliminated so that beds could be used for acute patients. Also involved were individuals and groups working in public and private welfare agencies. These included visiting nurses seeing more chronic patients as homecare programs expanded and an emerging group of public social workers developing a professional and organizational identity. Finally, there were civic leaders of various sorts and private philanthropic think tanks like the Milbank Memorial Trust. This collection of individuals and groups represented diverse values, motivations, and interests but all were responding to a series of new imperatives. These included political ideologies and values that made prevailing health and welfare arrangements seem both inhumane and inefficient and the rapid growth of cities that made existing institutions appear wildly inadequate in relation to need.

Closely connected with this heterogeneous movement whose epicenter was New York City was the wider hospital reform movement seeking to impose standardized institutional practices and infrastructures on the nation's potpourri of rapidly evolving hospitals. Much has been written about the quest for "efficiency" during the Progressive era and interwar decades. It involved efforts to standardize institutions and practices, but was also about bringing together diverse associations, agencies, and institutions, both public and private, in an attempt to coordinate and rationalize activities. Rational coordination also meant separating different categories of clients and providing each with appropriate care and services. This process had begun in the nineteenth century with measures like the transfer of the mentally ill from prisons and almshouses to asylums. By the early twentieth century, the match between clients and institutions remained incomplete and there seemed no lack of residual inappropriateness in existing arrangements. Many of those who were sick, disabled, and mentally ill remained in almshouses or other institutions where they did not receive medical care, in general acute-care hospitals that did not know quite what to do with them, or in private lodgings where no one cared for them. Efficiency demanded appropriate services for each condition.

The quest for efficiency was one of several links between reformers in the hospital-welfare sector and those in the public health field. There existed close ties at the local level where practitioners in the two domains often worked together. Nationally, the American Public Health Association (APHA) and the American Hospital Association (AHA) formed committees in the 1920s to study common problems, and in 1935 the AHA and newly created American Public Welfare Association (APWA) began doing the same. Representatives of each sector regularly attended and presented papers at one another's meetings. Both groups, inspired by a series of impressive early twentieth-century medical discoveries and innovations, were firmly convinced that diseases and conditions of disability that excluded many from the workforce had become remediable to some degree. Although rehabilitation of military veterans was not usually cited in this rhetoric—most probably because it occurred in separate Veterans Administration institutions subsidized by funds not available for the poor—the basic reform instinct seems to have been the same: transforming disabled charity or welfare cases into productive citizens capable of supporting themselves.[1]

Charitable work gradually evolved into social work where as early as 1913 spokesmen insisted, "the aim of charity is to supersede itself."[2] Such change involved a shift from caring for individuals in institutions to the apparently more economical aim of forestalling the need for institutionalized care.[3] As in the case of public health, such goals involved institutional reorganization and expansion into new realms of activity financed by government. "It is very clear that the old 'charities and corrections' have been transcended by the newer reasonable, democratic, constructive and preventive, as well as remedial, service to all the people within the state's domain. It is equally clear that the obligation to make good in these newer steps of progress rests alike upon formal government and civic community."[4] The new social welfare would minister to the socially deficient while preventing social deficiency. It would be decentralized, democratic, adequately organized, and standardized in its operations. Its proponents could also speak in tough-minded eugenic tones, rejecting "old ideals of false sympathy and race weakening charity" and seeking instead "the promotion of human adequacy."[5] Referring to New York City's Department of Public Welfare, which in 1922 had a staff of 4,200 employees and a budget of $7,370,550, one author insisted that it was "not at present the attitude of the department that it is dispensing charity but rather one of temporary aid and helpfulness to an individual who needs assistance to become an able-bodied and self-supporting member of society."[6]

Such aspirations were comparable to those of public health reformers. In 1923 the director of the Rockefeller Foundation pointed out that public health was "intimately related" to the field of public welfare and required the cooperation of all welfare agencies.[7] An account of the recent National Conference of Social Work that appeared in the *American Journal of Public Health* noted: "there was scarcely a section whose interrelations with public-health problems did not crop out at some point. The visiting sanitarians came away with a new realization that public health *is* social work of a very high order, and that social work that forgets or ignores the health factor is not worthy of the name."[8] It was this interaction between welfare and public health work that allowed chronic disease to emerge as a social problem after World War I.

Hospital and Welfare Surveys during the Interwar Era

In the early twentieth century, hospitals in American cities were managed in a variety of ways. There was the basic divide between public and private or voluntary institutions. In the public sphere, different municipal departments might manage different hospitals; for example, in the early twentieth century, New York City's municipally run hospitals were divided between the department of public health and the department of welfare. Such divisions made it difficult to confront daunting new challenges. Cities were growing rapidly, placing great strains on institutional resources. Such strains were exacerbated by the evolution of hospitals into fully medical institutions that sought to cure patients rather than merely to care for them. Voluntary hospitals had long sought to exclude chronic or dying patients on the grounds that beds could be better used for the curable.[9] Municipal hospitals had traditionally cared for the chronically ill poor, but now that acute care was becoming their chief preoccupation, doctors and administrators were eager to eliminate such patients. It is true that more and more diseases were considered treatable and thus appropriate for hospital care. But medical care for tuberculosis (and later cancer) sufferers, it was charged, was open only to those in the early stages of disease.[10] Since municipal welfare authorities administered many hospitals, discussion of hospital services often extended to welfare institutions generally.

Following World War I, surveys became the most common method for measuring a city's hospital resources and needs. In 1920 the city of Cleveland published a huge eleven-volume survey of municipal health resources that would become a model for other cities. It was directed by Haven Emerson, soon to be appointed professor of public health at Columbia University. The goal was to catalog existing facilities in order to determine present and future resource needs for a dozen or so conditions, including chronic diseases. Volume 10 of the survey was devoted to hospitals and dispensaries and reflected the views of hospital administrators. It complained that there were at all times "several hundred patients in the hospitals of Cleveland designed for acute cases, who are chronic cases and should not be in these hospitals at all. As a result, hospital service is rendered less available, and the acutely sick must often be denied needed care because beds are taken."[11] But chronic "cases" were not just a drain on the system; they, too, were not getting the care they needed. The shortage of dispensary and consultant services was especially harmful to ambulatory chronic patients. The survey concluded that municipal payments to institutions to care for chronic patients (rather than asking them to provide free care) would significantly improve the situation, while adequate social service departments would allow many individuals to be properly cared for in their homes. For the poor who could not remain at home, an existing infirmary should take over, at the city's expense, the role that general hospitals were no longer playing. Similarly, there was great need for sanatoria for the many tuberculosis patients now living at home.

Other cities also undertook hospital surveys, with many hiring Haven Emerson, who established himself as something of a survey industry. While a business model of efficiency certainly lay behind many of these surveys,[12] they were undertaken for a variety of practical reasons. In Cincinnati, a survey was organized to determine whether building campaigns launched by certain hospitals were justified.[13] In Buffalo, Emerson suggested, a survey was undertaken in order to deflect attacks from partisan politicians, while the goal in San Francisco was to decide how to allocate centralized funds in deficit. Other surveys reflected the desire of central charities like Community Chest for a clearer idea of existing facilities.[14] Such surveys frequently included short sections on the inadequacies of both chronic and convalescent care and emphasized the need for new and more appropriate institutions to provide needed medical services.[15] In New York City, the Public Health Section of the New York Academy of Medicine (NYAM) undertook in 1922 a hospital survey directed by E. H. Lewinski-Corwin, of the Academy and the Hospital Information Bureau of the United Hospital Fund of New York. The results were published in 1924.[16] Although this report contained an extensive section on convalescent care, there was only a short section on provision for chronic patients. Nonetheless, the report called for more facilities for chronic diseases and mental illnesses and the subject was again placed on the public agenda. Soon after, one of the men involved in the survey, Ernst Boas, a young cardiologist (and son of the famous anthropologist, Franz Boas) who since 1920 had been medical director of the Montefiore Hospital for Chronic Disease and was associate editor of the journal *Modern Hospital*, wrote to the head of the Public Health Committee of NYAM to suggest that the results of the recent survey underscored the need to undertake a new survey devoted specifically to chronic disease.[17]

The academy did not act on this suggestion, but elsewhere agencies undertook surveys devoted to chronic diseases among welfare populations. In 1925 the Pennsylvania Department of Welfare initiated a small survey of almshouses in a dozen counties.[18] Other cities undertook similar surveys, in some cases following the creation of a local coordinating committee bringing together a city's private and public welfare agencies. During the 1920s and 1930s, many such councils were set up with a few creating research departments to determine community resources and clarify a variety of welfare issues.[19] The Boston Council of Social Agencies was founded in December 1920. By the end of the first year, 176 agencies had joined. The council became interested in chronic disease because a project to build a municipal chronic-care hospital on the mainland to replace an existing one offshore on Long Island was abandoned in 1926 after construction had already begun. The council appointed a committee to examine existing care of the chronic sick and aged and to make recommendations. Haven Emerson was chosen to direct this survey.[20] The Cleveland Health Council organized a study in 1927 in which information was collected about each person who had a chronic disease known to any private or public agency.[21] In New York City, the Welfare Council of New York was founded in 1924 to coordinate the work of private and public welfare agencies. By 1929, seven hundred organizations were members. A research bureau was set up in 1925 and two years later began an ambitious program of empirical studies.[22] Chronic illness was one of the early subjects on its agenda and was placed under the direction of Mary C. Jarrett, who would later become a pioneering psychiatric social worker.[23] After publishing with Michael M. Davis a health inventory of the city that was critical of the health department for doing little about chronic or mental diseases, she went on to produce a study of care in private homes for the aged and then the results of a massive survey of chronic illness in the city.[24] In New Jersey, the state legislature passed a resolution in 1931 that resulted in a similar survey by the New Jersey Welfare Department.[25] In addition to formal surveys, numerous articles on the care of chronically ill people were published in journals like *Modern Hospital*, *Transactions of the American Hospital Association*, and *Public Health Nurse*. The most important publication was certainly a book by Ernst Boas and Nicholas Michelsohn published in 1929 and directed mainly at professionals in the field, which systematized the case for reforming and expanding chronic care.[26]

Welfare and the Problems of Chronic Care

Surveys established that many of those on welfare rolls suffered from chronic diseases or disabilities that frequently created the need for social assistance. The Boston survey, for instance, found that heart disease was the most common condition among welfare recipients, followed by neurological problems, cancer, and arthritis. Most were cared for in some institution, but there was little provision for the poor who were not destitute.[27] In Pennsylvania, it was reported that 90% of inmates of almshouses were infirm, chronically ill, or feeble minded.[28] The Cleveland survey found 460 chronically ill individuals, of which 398 were indigent.[29] In New Jersey, it was claimed that among the 60,000 to 75,000 individuals under care of social or medical agencies, approximately 20,000 were chronically ill.[30] In her study of New York City, Mary Jarrett found more than 20,000 people incapacitated by injuries, handicaps, or disease. She emphasized that large numbers of chronically ill and disabled individuals were not old and many were in fact children. She estimated that about 10% of the total population of the city suffered from such conditions, with nearly 10% of these wholly disabled.[31] Such surveys inevitably concluded that chronically ill people and invalids cost society a great deal of money, not to mention lost productivity and work.

Making matters worse, institutions to care for such individuals were grossly inadequate in at least two respects. There were not nearly enough of them and those that existed were of the wrong type. Where were these chronically ill persons and invalids located? Few were in private hospitals, which had systematically excluded "incurables" during the nineteenth century and continued to do so. There were a number of inadequate choices available, including old-age homes and municipal infirmaries. But three sorts of institutions played a particularly significant role: almshouses, general hospitals, and chronic-care hospitals.

Almshouses

The traditional basis of residential care in the United States as elsewhere was the poorhouse, or almshouse, which had emerged during the first half of the nineteenth century to house (and some would say punish) the poor as well as those who were sick, insane, and mentally deficient. Conditions were grim in order to encourage employment or familial care. Almshouses gradually became distinguished from other sorts of residential institutions like prisons, orphanages, and asylums, while the poor health of large numbers of indigent people led to the provision of medical care in many institutions and sometimes to their gradual transformation into general public hospitals. Nonetheless, by the early twentieth century, almshouses remained a significant part of the welfare sector. Providing primitive custodial care, they were unpopular institutions and objects of intense criticism. They were identified with poor living standards and seriously inadequate medical care.[32] They remained, in Charles Rosenberg's felicitous phrase "at once refuge and punishment for the morally and physically incapacitated."[33] In even stronger tones, M. Holstein and T. R. Cole write: "The poorhouse became a symbol of brutality and corruption. As public sentiment became increasingly indifferent to the poor, the almshouses and their inhabitants were, in essence, abandoned."[34]

Reformers sought to eliminate the stigma associated with poverty relief. In 1903 New York's commissioner of charities tried to rename the city's almshouse the Home for the Aged and Infirm on the grounds that "inmates of almshouses were more nearly related to hospital patients than paupers."[35] Nursing reformers like Lavinia Dock had advocated nursing care in almshouses without much success.[36] But the status of these institutions worsened. It was generally believed that the transfer of entire categories of clients (patients with tuberculosis, mental illnesses or handicaps, orphans) to specialized institutions had produced a clientele increasingly made up of the elderly poor, further increasing the marginal status of almshouses.[37] But reality was more complex. A report issued by Women's Department of the National Civic Federation argued that the almshouse was not a refuge for old age but for feeble-mindedness and chronic disease. A survey of seventy-five institutions in four states designed to study the problem of old age dependency found that almost half of inmates were less than 65 years of age and advocated the "hospitalization of almshouses to take care of chronic sick"; at the same time, it reported appalling conditions and understaffing.[38] Ernst Boas, too, saw the problem of the almshouse as one of chronic disease rather than age, with only a small proportion (7%) of the estimated 85,000 inmates nationally in 1923 able-bodied and the majority (55%) totally incapacitated. He reserved some of his sharpest criticism for almshouses where sick diabetics who happened to be poor were deprived of life-saving insulin; for Boas, this deprivation epitomized denial to the poor of all the advances of modern medicine. There was some possibility of change. Boas cited the case of Luzerne County Almshouse, in Pennsylvania, which had been transformed into a medical institution and changed its name to the Retreat Home and Hospital for Chronic Diseases.[39] But for the most part he did not see almshouse reform as a viable goal.

General Hospitals

By the early twentieth century, some general hospitals like the one in Philadelphia remained "a hybrid of hospital and almshouse" characterized by "vagueness of distinction between dependence, sickness, and delinquency."[40] Nonetheless, many were in the process of becoming medical institutions focused on the provision of acute care. Thus, the presence of patients who did not require active medical treatment or whose situation appeared hopeless seemed increasingly disturbing. Haven Emerson reported that in Louisville 12% of hospital beds were filled from 1 to 3 months and 11.3% from 3 months to 10 years.[41] The presence of elderly and chronically ill people led to the overcrowding of hospital wards and interfered with and sometimes prevented the treatment of acute-care patients. "Chronics," it was charged, depressed younger patients, were not good teaching material, and hampered efforts of general hospitals to be accepted into mainstream medical education. Worst of all, hospitals were becoming increasingly expensive places to care for such patients. Emerson estimated that they could be cared for at half the cost or less with far better results in more suitable institutions.[42] Some argued for the creation of special wards for the chronically ill; others found that where such wards had been established, they tended to be ignored by the staff and sank "into a custodial lethargy."[43] As hospitals became more and more defined by acute illness and timely therapeutic intervention—in 1934 Bayview Asylum in Baltimore was officially renamed Baltimore City Hospital although it remained under the control of the department of welfare until 1964[44]—the disconnect between ideals and the reality of increasing numbers of chronically ill patients became harder to bear. The solution, in Emerson's words, was that "the convalescent home and the home for chronics and incurables are necessary to supplement the best type of acute general hospital service."[45]

Chronic-Care Hospitals

Many cities had private institutions for specific sorts of long-term patients. Boston, for instance, had seven such institutions, the earliest one, the Channing Home for the Treatment of Incurable Diseases, opened in 1857 with fifteen patients. Caring for chronic patients was a motive in the founding of Philadelphia's Episcopal Hospital and New York's St. Luke's Hospital.[46] Such chronic institutions were usually small, however, and were frequently located in out-of-the-way places—on nearby islands as in Boston and New York, or in remote suburbs, as in Philadelphia, Chicago, and Washington. Doctors in training avoided them. Wards were usually crowded, poorly kept, and meagerly supplied; staff doctors were often medical hacks, ill trained, uninspired, and uninspiring.[47] Waiting lists for admission were long, and only custodial care was available in most until the 1940s. No one denied the need for custodial institutions, "but what is urgently needed in addition are hospitals where chronically but not hopelessly ill patients can be salvaged and reclaimed—institutions similar to the sanatoria for the treatment of tuberculosis."[48] A central task of these new institutions would be clinical research into chronic diseases.

There existed only a few examples of such institutions. First among them both chronologically and by reputation was the Montefiore Hospital for Chronic Diseases established in the Bronx by a group of wealthy Jewish philanthropists in 1885. There was in the early years tension between the benefactors, who saw Montefiore in traditional terms as a custodial home, and its chief physician, Dr. Simon Baruch (father of the financier Bernard Baruch), who insisted that it was a medical institution devoted to treatment and research. Disagreements arose about whether it should be called a "home" or a "hospital" and various compromises involving both terms were made. In 1914, as a new building in the Bronx was being opened, the medical director, Dr. Siegfried Wachsmann, proposed to eliminate the word *home* from the awkward name of the institution (Montefiore Home and Hospital for Chronic Invalids and Country Sanitarium for Consumptives) and call it Montefiore Hospital for Chronic Diseases and Country Sanitarium on the grounds that "hospital work . . . is of much greater prominence than that of the nursing home. Research laboratories are maintained, and the institution is much like other hospitals, except that emergency cases are not handled."[49] The attachment to tradition persisted, however, and Wachsmann's suggested name was rejected in favor of Montefiore Home, Hospital, and County Sanitarium for Chronic Diseases.

By the 1930s, the status of the institution had been resolved. Affiliated with Columbia University, Montefiore was a model hospital of a new type. With 742 beds in its Bronx building and another 250 in its Westchester annex, it admitted a wide variety of patients, including tuberculosis patients lodged in a separate building, crippled children, and sufferers of venereal disease. In 1922 Ernst Boas opened a new training school for nurses, arguing that chronic care required special training. Montefiore provided rehabilitation services, occupational therapy, and advanced medical treatments. It was one of the first institutions to use adrenalin in the treatment of asthma. It was involved in therapeutic research and was one of the testing hospitals for the use of insulin in diabetic patients.[50] When Lee Frankel of Metropolitan Life insisted on more research into chronic diseases, he emphasized the need for more "hospitals of the Montefiore type."[51] As part of the celebration of its fiftieth anniversary in 1934, the hospital organized what was described as the "first clinical conference in medical history devoted to chronic diseases."[52]

As a medical and research-oriented facility, Montefiore could try to move beyond its traditional indigent clientele. In 1912 four of its wealthiest directors collectively donated $200,000 to build a pavilion for private patients. "The middle class and the wealthy are subjected to the same ills and sufferings as the poorer classes, but it is only the very wealthy who can ward off such discomforts as can be entirely eliminated only by money, that is, where space, skilled nurses, and physicians can be obtained regardless of the cost." While fees would pay for much of the costs, it was hoped that the new pavilion would also attract further donations from friends and relatives of patients.[53]

Another chronic-care institution on the new model was the privately endowed Robert B. Brigham Hospital for Chronic Diseases opened in 1915 "for the care and support, and medical and surgical treatment of those citizens of Boston who are without the necessary means of support and are incapable of obtaining a comfortable livelihood by reason of chronic or incurable disease or permanent physical disability."[54] Although it was open to patients who had many chronic conditions, it specialized in orthopedic treatment. The hospital was established "on the striking foundation that the treatment of chronic disease is a problem of applied physiology, and that chronic disease can be successfully treated." The founders might, in the words of one of its leading lights, "have adopted the hopeless anatomical view of disease and founded a home in which decrepit men and women could receive shelter and food, and pass their declining days in peace—a dreary congeries of helpless and hopeless misery with death as its goal." Instead, they chose to create an institution where "help and hope were to be the guides and an efficient life the goal." This meant in practice that it was organized differently from other hospitals. Rather than being divided into services based on special treatments for different kinds of anatomical lesions, there was a single service that focused on functional restoration rather than anatomical integrity of organs. Research was a central function justified by a reversal of the customary reasoning: the long duration of the diseases treated made the hospital an especially fertile place for clinical research.[55]

There were a few other examples of modern chronic-care institutions. In 1925 Alameda County Hospital in California opened special wards. These were reported to be airy and spacious, with solariums and special staff whose goal was rehabilitation and making patients' lives more pleasant. Occupational therapy was emphasized. Above all, money was being saved. "By segregating chronic patients, and with careful planning of the wards and organization, it has been possible to give the chronic patient those things which he needs at a cost much less than that in acute wards."[56] A year later, these wards had been transformed into Fairmont Hospital, providing indigent elderly and disabled patients with medical and surgical care as well as a pleasant living environment. Westchester County began with an infirmary for chronic illness under the supervision of the county general hospital. By 1928, it was planning a large hospital. Several years later, Cincinnati had built the first unit of what was to become a county chronic-care hospital to replace the city almshouse.[57] Not all experiments were successful. Michael Reese Hospital in Chicago tried to fill empty beds and increase income by opening a division in the private patient building for individuals suffering from chronic illness (excluding tuberculosis and mental disease). In the midst of the Depression, this experiment was not successful as families struggled to reduce expenses. The superintendent of Michael Reese concluded: "It is my belief that at this time general hospitals should not expend any funds upon the development of a program to attract private chronic patients."[58]

Ernst Boas and the Rational Organization of Chronic Care

The person who undoubtedly did most to advance the cause of chronic-care hospitals in New York City was Ernst Boas. The book he co-authored with Nicholas Michelsohn and published in 1929 took up and elaborated on a number of themes he had developed in numerous articles since 1920; its publication made him the city's leading expert on the chronic disease problem. Like others, Boas cited the rise in mortality from chronic diseases (and the decline in infectious diseases)[59] but this was not at the heart of his commitment to chronic care. More central was his belief that many illnesses ordinarily considered incurable could in fact be cured, ameliorated, or at least stabilized if healthcare was rationally organized. Although tuberculosis and insanity were clearly chronic diseases, Boas excluded them from discussion because the principle of community care for those afflicted had already been accepted in the form of sanitaria and asylums. Consequently, he utilized a limited definition of chronic disease:

> chronic physical disability is determined largely by diseases of the heart and arteries, organic affections of the nervous system, cancer, non-tuberculous diseases of the lungs, the various forms of rheumatism and Bright's disease, by diabetes mellitus and other disturbances of the glands of internal secretion or of metabolism . . . Physical incapacity arising from these diseases is at first insignificant but gradually assumes ever greater proportions. There are no cures for most of these affections, but in many instances the progress of the disease can be arrested or retarded and the patient can be granted years of useful life.[60]

The last point was critical because it rejected the traditional association of chronicity with hopelessness or the inevitable consequences of old age. Boas told the members of the American Hospital Association in 1927 that the word *incurable* should be removed from the dictionary.[61] "Senile decay" only occurred in Boas's view after the age of 70. "Persons between their fiftieth and seventieth years, who are disabled or infirm, should be regarded as sick not as suffering from the natural decrepitude of old age."[62] In fact, half of the patients in Montefiore Hospital were under the age of 50. In the next few years, welfare surveys would confirm that chronicity was not just a problem of the elderly.

Boas developed a system of classification that became commonplace in discussions of the chronic disease problem.[63] There were first of all the able-bodied who could care for themselves at home and could be best served by a pension or financial subsidy. Boas, however, showed little interest in this group.[64] He was most concerned with individuals needing institutional care, which he divided into three groups. Class A was composed of those requiring sophisticated medical diagnosis and treatment; these included people who had heart disease or stroke patients who could achieve partial rehabilitation. Class B was made up of those needing skilled nursing care and little else because rehabilitation was not possible. Class C comprised those who needed only custodial care such as assistance in dressing, bathing, or eating. Each of these classes required very different kinds of facilities. Because health status changed, patients might move from one facility to another and back. It was thus a good idea for the different institutions to be in close physical proximity to one another.

To the extent that he occasionally discussed the causes of chronic illness, Boas continued to express a social perspective that was becoming less common in public health circles. In a book that he wrote in 1940, he briefly and according to a review in *JAMA* "rather impetuously" included the following: "An aroused community, in self preservation as well as for humanitarian reasons, must face the fact that the bare existence forced on so large a proportion of our population generates and accelerates disease and undermines the health of vast numbers of citizens. Steady employment, fair wages and proper housing will prevent more disease than most laboratory research and educational ballyhoo about disease prevention."[65] Boas's main concern, however, was care for the already ill. He had no objection to special chronic-care wards in general hospitals but he usually emphasized special hospitals, with New York City's new Cancer Institute as a model that could be applied to all chronic disease. Almshouses might be transformed into such institutions by investing enough money for proper equipment and medical care. Or county hospitals might build a special pavilion for chronic patients on their grounds. Yet another possibility was for the state to build a number of large chronic-care hospitals, analogous to state hospitals for the insane or tuberculosis sanatoria, which should be close to large cities and associated with medical schools, at once raising medical standards and providing material for teaching and research. Rehabilitation was a critical function and such hospitals required, in addition to the latest medical technology, departments of physiotherapy, occupational therapy, and social work, as well as facilities for arts and crafts important for morale. Likewise, they needed to be pleasant and livable places with landscaping, "good southern exposure and protection from the wind."[66] Chronic care also demanded special nursing skills, in no way inferior to those needed in acute-care institutions, as the director of nursing at Montefiore reminded readers of *Modern Hospital*.[67] During the next decade, the city of New York would, in no small measure due to Boas's influence, create a new kind of chronic-care hospital.

I ❤ New York

Interest in chronic disease developed early in New York City, agitating the New York City Health Department and the NYAM during the first decades of the twentieth century (see chapter 1). Throughout the 1920s, this interest persisted. The New York City Department of Public Welfare in 1924 opened the first municipal cancer clinic in the United States. The clinic, aimed at those who could not afford care in private hospitals, provided walk-in diagnosis and free treatment; hospital facilities were available for two hundred patients at City Hospital on Welfare Island.[68] In that same year, following publication of the NYAM hospital survey, a letter from Ernst Boas prompted representatives of the NYAM to meet with Bird Coler, the city's first comptroller, in an unsuccessful attempt to transform one or both of the hospitals on Welfare Island into a chronic disease hospital.[69]

Two developments were critical to the intensification of the chronic disease problem in the city. First, the Welfare Council of New York founded in 1924 became a center for empirical studies and reform ideas. Second, the division of the city hospitals between the department of welfare and the department of health came to an end in 1928, when a single department of hospitals took charge of all municipal hospitals.[70] This produced a central locus of power to which calls for hospital reform could be addressed and from which they could emanate.

Columbia University in 1932 began offering eight-week extension courses for doctors on chronic diseases. More significant was the major survey of chronic disease in the city published in 1933 by Mary Jarrett.[71] This survey was far more exhaustive than those of other cities and initiated a new program to deal with the chronic disease problem, which appeared to reformers to have intensified as the Depression aggravated the welfare burdens of the city. Fiorello La Guardia was elected mayor of the city in 1934 on a reformist platform. He appointed as his new commissioner of hospitals S. S. Goldwater, formerly medical director of Mount Sinai Hospital and long concerned with chronic disease (see chapter 1). The La Guardia administration placed considerable emphasis on the improvement of healthcare institutions. On the national scene, Franklin D. Roosevelt was elected president of the United States in 1933. The New Deal was accompanied by significant pressure for comprehensive healthcare reform and the move of influential New Yorkers to Washington, DC, which extended the problem of chronic disease from the local to the national political stage. I discuss the national scene in detail in the following chapter. Here I merely note that the considerable dynamism around health reform at the national level helped fuel New York City's vigorous response to the chronic disease issue.

The publication of Jarrett's chronic disease survey in 1933 was a critical turning point. Among surveys of this type, it was unique in the resources and research that had gone into it. Ernst Boas wrote the preface and the Research Committee included Haven Emerson and Louis Dublin. The statistical bureau of Metropolitan Life provided technical aid, while the Rockefeller Foundation, the Commonwealth Fund, the New York Foundation, and the Macy Foundation helped with financial grants. In addition to the work of Boas, Jarrett also drew on the public health literature on the subject (especially the Massachusetts study of chronic disease that had recently been published), taking it for granted that public health authorities had assumed responsibility for cancer and were abandoning the distinction between prevention and treatment. After documenting the seriousness of the problem, Jarrett proposed a comprehensive community program to deal with it. Among the many aspects of the program were the following.

1. To create public awareness of the seriousness of chronic disease "with [its] resultant disability, suffering, and economic loss . . . and to arouse the sentiment that the responsibility of society is at least as great for the chronically ill as for the acutely ill . . . In the control of chronic disease, prevention and cure are two aspects of the problem rather than separate divisions of work."
2. To integrate mental health services with medical services in the study and treatment of chronic disease since psychic factors play a large role in the causation and continuance of chronic illness.
3. "To study the extent to which social and economic factors contribute to chronic disease and the possibilities of preventing disability through social services."
4. To promote public responsibility for the care of the chronically ill as is now the case for tuberculosis and mental disease.
5. To establish a municipal chronic-care hospital "administered and equipped as a modern medical institution" and actively pursuing medical education and research.
6. To form a central planning committee to define the role of the health department and private agencies, to organize the community's resources, and to initiate the development of additional facilities.[72]

To implement such a program, the Welfare Council of New York in the spring of 1933 set up a Committee on Chronic Disease, "endeavoring to promote the more adequate care of the chronic sick by all of the many agencies which concern themselves with this problem."[73] The chair of this committee was Ernst Boas, now recognized as the city's leading authority on this subject. As *the New York Times* put it some months later, the committee's goal was to establish a program "to correct the present chaotic situation affecting the chronically ill part of New York's population."[74] Shortly thereafter, a subcommittee drafted a policy memorandum whose main point was the construction of a new five hundred–bed hospital for chronic disease. Approved by the Committee on Chronic Illness and the Executive Committee of the Welfare Council, the report was sent by Boas for approval to the Committee on Public Health Relations of the NYAM and to the United Hospital Fund so that a joint recommendation urging a general plan for the care of the chronically ill could be submitted to the incoming city administration.[75]

Somewhat surprisingly, E. M. Bluestone, Boas's successor as medical director of the venerable Montefiore Hospital, was one of the few voices raised against this proposal, which he attacked regularly during the next five years. There may have been some personal pique involved; Bluestone pointed out that the backers of the new hospital talked about deficiencies of policy without mentioning the successes of Montefiore Hospital. But his substantive argument was that segregating chronic patients in an island hospital was a form of "dumping." It separated them geographically from the population and detached them medically, thus preventing the integration of chronic illness into mainstream medicine. Nor were such hospitals economical if organized properly because they had to duplicate the technological resources of general hospitals. In contrast, Bluestone argued, if chronic patients needed less intensive care than acute patients, as was suggested, keeping them in special wards of general hospitals would actually reduce per capita hospital costs, especially in view of the fact that many hospital beds and wards remained empty. Furthermore, such wards would make chronic illness visible to the elite scientific physicians who worked in general hospitals. Rather than building new hospitals, money could be better spent by maintaining patients in their own or residential homes. Like others, he believed that providing small disability pensions would remove large number of chronic and custodial patients from hospitals and allow them "into homes where they could be among their own" and looked after by subsidized practitioners and visiting staff from general hospitals.

In arguing that general hospitals could do more for research than segregated hospitals, Bluestone pointed out that apart from his own Montefiore Hospital, no chronic institution had made any real contribution to treatment and that this was likely to remain the case. "You cannot expect to recruit a group of super-physicians to do in an isolated and segregated institution for the chronic sick what the high grade visiting staff in the acute general hospital is unable or unwilling to do for them." He insisted that dumping chronic patients in segregated hospitals was to allow general hospitals to disregard their moral obligation to solve the dilemmas of chronic disease. The challenge of chronic disease should remain "where all can see it." There was, of course, no question that fully custodial patients should be removed from general hospitals. Yet medical scientists should not ignore those chronic patients who "yield slowly to the efforts of the medical scientist and to the efforts of Nature." Bluestone recommended a detailed survey of conditions in public chronic-care hospitals in order to improve them and prepare for integrating them with acute general hospitals "in order that their inmates may not be forgotten." He closed his first missive with the statement that a new hospital on Welfare Island "is unworthy of the medical profession and of the profession of social service that is so closely associated with it."[76]

Some of Bluestone's suggestions, like doing a thorough study of resources for chronic patients or giving municipal subsidies to hospitals providing chronic care, were in fact followed. But on the main issue he remained a lone voice. Building a new hospital became the visible sign of the city's commitment to solving its chronic disease problem. In May 1934 the Committee on Chronic Illness sent recommendations to the city's commissioner of hospitals, S. S. Goldwater, for an integrated community program of chronic care and including a proposal for one modern chronic disease hospital on Welfare Island and for a second one in Brooklyn. The NYAM, the United Hospital Fund, the New York City Visiting Committee of the State Charities Aid Association, and the Executive Committee of the Welfare Council of New York City endorsed the proposal. Goldwater and Mayor La Guardia both responded enthusiastically. La Guardia immediately requested federal funds to build two huge modern hospitals for chronic diseases on Welfare Island, each containing 1,500 beds.[77] He received only partial federal funding and thus focused on building a single hospital. A second hospital was planned for some time in the near future but World War II delayed the opening of the Bird Coler Hospital—also on Welfare Island—until 1953. Meanwhile, Goldwater presented a national radio address in 1935, the text of which was published in *Modern Hospital* and reprints of which were distributed by the Welfare Council of New York.[78] His message was largely based on the report of the Committee on Chronic

Illness. He began with the melodramatic declaration typical of the decade that the growing rate of chronic disease "suggests that America may some day become essentially a nation of invalids." He especially emphasized the growing problem of mental illness that was filling the nation's asylums. The solution was hospitalization and cure in new types of chronic disease hospitals in which "the cases could be sorted out and classified so as to facilitate close observation and study." And while private philanthropy could help, "the problem is so vast that nothing less than a government sponsored program will suffice." He thus presented the comprehensive program developed by the Welfare Council and the NYAM as a model and asked for "its consideration by every city in the country which accepts the principle of public responsibility for the care of its sick."[79] He offered the aid and cooperation of the New York City Department of Hospitals to local officials in other cities interested in dealing with the problem.

Despite the appeal to the nation, one suspects that Goldwater's main audience was local. The Welfare Island hospital was the central and most visible element of New York's chronic disease strategy. It received only limited federal financial aid, but groundbreaking took place in 1937 at the site of the recently demolished penitentiary. Budgeted at a cost of $7 million, it was billed as "the largest and most modern hospital in the world devoted exclusively to the scientific care of chronic disease." Two years later, the hospital, now called the Goldwater Hospital, was completed at a cost of almost $8 million.[80] Although it was projected to have 1,600 beds, it began with about 1,100. It comprised five buildings in chevron shape to give maximum sunshine and unobstructed views of the river and city. All patient rooms had a southern exposure. The New Deal's Works Progress Administration (WPA) funded the production of abstract mural paintings for some of the common rooms. Chronic patients were to be brought in from other city hospitals along with about two hundred nurses. The only chronic diseases excluded from the hospital were mental illnesses and tuberculosis; these patients were treated in more specialized public institutions. Research into chronic disease was to be central to its mandate.

Goldwater Hospital was the crowning jewel of New York City's chronic health agenda. In the mid-1930s the WPA financed two innovative municipal programs. The first program sent visiting housekeepers to the homes of elderly people. The second sent them to the homes of chronically ill individuals. A visiting nurse service and supervision by physicians were part of this program. One aim of the project was to determine whether and to what extent "it is better for the patient and cheaper for the city to care for dependents chronically ill in their homes or in institutions."[81] Private homes providing custodial care for elderly and chronically ill people had long existed in the city. In 1936 the city had about 14,708 beds in such private homes. This was roughly equal to the number in county- and city-administered homes. The Social Security Act of 1935, by providing assistance to elderly and unemployed Americans, expanded the potential clientele of these homes, both by increasing the number able to pay and making *only* those in private homes eligible for this financial assistance. In 1938 the Welfare Council of New York developed strict standards for such custodial care, distinguishing between normal homes for the aged and those wishing to serve as hospitals for the chronically ill, based on facilities and types of nursing and medical care provided. The hope was that this would add from five thousand to eight thousand new, mainly custodial beds for the city's chronically ill residents.[82]

Perhaps the most interesting initiative by Goldwater was his effort, supported by the deans of the five medical schools in the city, to create a center for research on chronic disease to be located temporarily at the Metropolitan Hospital on Welfare Island until the new chronic disease hospital was completed. Each medical school would manage a unit of the hospital. Only in this manner, it was argued, could new methods be discovered to cure or alleviate chronic illnesses. Starting in 1935, Goldwater tried to collect $25,000 from private individuals and agencies to encourage the city council to provide funds, but the task proved challenging. Metropolitan Life Insurance Company (MLI), which had promised $5,000 if $20,000 more was collected, contributed only $3,000 since the promised amount was not raised.[83] Nonetheless, Goldwater persisted and managed to get the city to provide funding. In 1936 David Seegal, professor of medicine at Columbia University, was appointed director of the Research Division of Chronic Diseases in the New York City Department of Hospitals at an annual salary of $8,000. MLI refused to donate more money on the grounds that these "projects, like these diseases, are likely to be long drawn out and chronic" and were unlikely to affect either the company or its policyholders.[84] But the department of hospitals contributed a modest sum annually, and in 1938 the research center received a grant of $66,000 from the Rockefeller Foundation that assured its continued existence. The prestigious journal *Science* published an article on the research being carried out by seven individuals or groups in this division.[85] This included work on cholesterol metabolism and arteriosclerosis, effects of hypertension on kidneys, and psychosomatic studies of rheumatism and arthritis. By the time the Goldwater Hospital was completed, there were plans for three research units with at least ten physicians each.

Meanwhile, efforts to educate the population about chronic disease continued. In 1939 the Welfare Council of New York City launched a public relations campaign using Goldwater's phrase "America may some day become essentially a nation of invalids."[86] The *New York Times* reported in July 1939 that "a diorama was set in motion at 6 o'clock last night on the balcony at the Grand Central Station. The large rectangular structure is about six feet high and has three windows, before which curtains are raised and lowered, showing, respectively conditions in the past, present and future with regard to such diseases."[87] It had been financed by the "Built by the WPA Art Project" on behalf of the Welfare Council and Mary Jarrett set the diorama in motion. After two weeks the diorama began to circulate around health centers throughout the city. Grand Central Station seems to have been a favorite of the Welfare Council. A few days later, it placed an exhibition there, depicting "a model of "A Chronic Disease Hospital and Health Center of Tomorrow," graphically illustrating how a community must plan the control of chronic diseases for the young, the middle aged and the aged . . . The model shows a health center from which will radiate public health activities, such as health education; consultation clinics; visiting doctors, nurses, social workers and housekeepers, and housing, factory and school inspection."[88] The *New York Times* was enthusiastic about the integrated system that this exhibition presented:

> For young, schools with medical services, classes for handicapped, rheumatic fever sanatoriums, foster homes for underprivileged children and light airy spacious and clean tenements. For middle-aged, there would be a chronic disease hospital with occupational therapy, an institutional home for the handicapped and crippled, a visiting housekeeper service, medical care at home by visiting nurse and physician, and care at a private doctor's office for ambulant cases. For the aged there is suggested an institutional home, a research laboratory for the causes of aging, and boarding homes for the aged.

This exhibition as well went on to tour the city.

Conclusion

New York City was in numerous ways unique. But by the end of the 1930s, the healthcare of indigent people suffering from chronic diseases was well on its way to being transformed from a local to a national issue. The AHA had long been concerned about the chronic patients taking up precious beds. But by the end of the decade some of its members were asking if hospitals did not have a significant role to play in chronic care, so long as new resources were made available. In 1930 the APWA was founded initially for state welfare officials. It was soon opened to professional staff members of public agencies at the federal and local levels as well.[89] Like members of other associations devoted to hospitals and/or welfare, those in the AHA and APWA had heard papers at annual meetings and had published articles in their respective journals on the chronic disease problem during the 1930s. But in the last years of the decade they began to intervene more directly. The APWA sponsored a report in 1937 on the relationship between welfare agencies and hospitals.[90] Three years later, a joint commission of the APWA and AHA produced guidelines for the chronic-care hospitals and nursing homes that were cropping up as a result of the transformation of almshouses into medical or semi-medical institutions.[91] Providing momentum to the spread of welfare and public health activism around chronic disease was the New Deal political campaign to transform healthcare in America.

CHAPTER FOUR

New Deal Politics and the National Health Survey

During the 1930s, chronic disease became widely recognized as a critical social problem in the United States. One reason was the growing influence of the cancer movement, then attracting considerable financial support and public attention. Cancer had a life of its own, however, and the intense fears, hopes, and activities surrounding it only occasionally extended to other chronic diseases. A more compelling cause was the move to Washington of many New Yorkers who accompanied Franklin Roosevelt to the nation's capital and who brought with them the city's concern for chronic disease. One of these, Harry Hopkins, who had been firmly established in New York City's welfare and public health scenes (in 1928 he had been part of the commission that reformed the city's health department), was in a position to allocate huge sums of money. One of the projects he funded was the National Health Survey on Chronic Disease and Disability (NHS) of 1935–36. It was the largest morbidity survey undertaken up until that time and firmly embedded the notion of a chronic disease crisis in the American political landscape.

Why was a study of such unprecedented scope undertaken at this time? I argue that it was part of the larger strategy to reform healthcare in the United States that had been going on for more than a decade. The answer to a second question—why a morbidity study?—has to do with the way falling mortality rates could be used as an argument against health reform and the need to counter that argument. A third question—why the focus on "chronic disease" and disability?—demands a more complex answer. Concern with the problem of chronic disease had intensified during the previous decade. But it was not yet a major issue for most health reformers, including those who initiated the survey. Although the focus of the NHS on chronic disease was part of the campaign to reform American healthcare, there was no inherent reason for such reform to emphasize chronic diseases.

European nations like France introduced health insurance during this period without invoking a chronic disease problem. That events unfolded in the United States in this particular way reflected to some degree the influence of movements for periodic health examinations, expanded public health, and welfare reform that all emphasized chronic disease and had a cumulative effect absent in Europe. But the primary reason that chronic disease became so prominent was that leading American health reformers created new kinds of data to support their position and, along the way, took the field of epidemiology in innovative directions. In seeking to show that terrible health among Americans demanded radical reform, they brought together the concerns and knowledge-gathering methods of public health epidemiologists and welfare reformers and gave new meaning and urgency to the concept of "chronic disease." The survey was then utilized by a New Deal interdepartmental committee seeking to coordinate national public health and welfare programs. This committee further contributed to advancing the chronic disease agenda.

Cancer in the 1930s

The cancer movement was in a class by itself when it came to gathering support and collecting money; it provided a model for advocacy groups around heart disease, polio, and other conditions. The more general problem of "chronic" or "degenerative" disease had emerged in different social spaces and attracted little of the public attention or extravagant private donations earmarked for cancer. An exception was the $1 million gift that the philanthropist Albert Lasker and his first wife donated in 1928 to the University of Chicago to support a research foundation devoted to "degenerative" diseases.[1] The source of their motivation is not clear, but newspaper reports about the donation discussed the possibility of extending lifespan, suggesting the influence of the Life Extension Institute (LEI) (especially clear in the *New York Times* report). But this first Lasker donation was the exception that proves the rule since the Lasker name eventually became closely associated with the cancer movement.

The establishment of a cancer program in Massachusetts in 1927, the earlier development of a municipal cancer clinic in New York City, and the creation of public programs in six other states by 1940 give some indication of the continuing impact of this disease on the American psyche.[2] Although legislative efforts in the late 1920s to promote more government involvement in cancer research were not very coherent or successful—offering, for instance, financial rewards for a cancer cure—private initiative did much to advance the American cancer agenda. The cancer lobby expanded while popular media called public attention to the disease and fostered mass participation. In 1936 the American Society for the Control of Cancer formed a Women's Field Army to educate women about the prevention, detection, and treatment of cancer. Cancer clinics, hospitals, and wards multiplied to such an extent that the American College of Surgeons (ACS) published guidelines for setting up cancer units and accredited them. The number of such clinics and hospitals increased from 13 in the early 1920s to 490 in 1940, with 345 receiving ACS approval. (In Canada, much influenced by the US movement, three provinces had by the 1930s set up public cancer programs.)[3] Growing faith in the power of science brought philanthropists into the picture. In 1936 John D. Rockefeller donated $3 million and valuable land for the rebuilding of the Memorial Cancer Hospital in New York City. In 1937 another wealthy New Yorker provided Yale University with a $10 million endowment "devoted primarily to medical research into the causes and origins of cancer."[4] Major cancer institutes were established in Philadelphia (1932) and Chicago (1937). If tuberculosis provided the initial model for a broad social movement responding to a disease, cancer offered another model from the 1930s on. It was now being treated in specialized wards and hospitals and was the subject of considerable research. While not quite equaling tuberculosis and mental illness as a recognized communal problem demanding state intervention, cancer nonetheless received substantial public attention.

Gradually the federal government also became involved. The National Institute of Health (NIH) set up in 1930 had pursued an active program of research on cancer since 1922.[5] The Social Security Act of 1935 allocated more funding to the research activities of the NIH, with work on cancer remaining at the forefront. During the next few years, pressure from legislators became more focused and more effective, energized by the larger battles for healthcare reform. The crowning achievement occurred in 1937, when Congress unanimously passed a bill creating the National Cancer Institute (NCI). In addition to an allocation of $750,000 to build a laboratory, an annual appropriation of $700,000 was included in the law. For the first time Congress authorized an agency of the Public Health Service (PHS) to make grants to outside researchers and institutions and to create a program of training fellowships. A syphilis-control campaign initiated by the surgeon-general, Thomas Parran, was in full swing by the fall of 1936, enabling Parran to turn his attention to cancer. By 1937, the media were discussing the PHS's effort to "conquer" cancer and a national survey on cancer incidence was underway. While not itself providing direct treatment of cancer patients, the NCI was authorized to buy and lend radium to therapeutic institutions.[6]

The successes of the American cancer movement were striking. Nonetheless, they did not lead directly to greater interest in the wider chronic disease problem. Cancer had its own constituency and during these same years Britain and France developed national programs of cancer therapy and prevention without any reference to chronic disease. If anything, the results of the National Health Survey on Chronic Disease and Disability were used to justify after the fact the huge investments being made in the NCI. Nonetheless, the cancer act established a pattern whereby the emphasis on combating chronic diseases would focus largely on research. This emphasis intensified when the NCI soon abandoned the purchase of new radium, its only practical therapeutic activity. The organization of research and prevention efforts around specific disease categories would continue to be the dominant pattern. In 1938 the PHS announced that it was encouraging cancer control programs and offered consultation services to interested states.[7] By then, a growing consensus about the seriousness of a wider chronic disease problem was emerging, due in large measure to the National Health Survey.

Health Politics and the Origins of the Survey

The NHS of 1935–36 was technically not the first national morbidity survey undertaken in the United States. John Shaw Billings used the censuses of 1880 and 1890 to collect data about sickness incidence. But his controversial effort was largely unsuccessful, as we saw in the introduction. In contrast, the well-financed NHS received wide popular support and was organized with industrial efficiency.[8] About 2,800,000 people in 19 states were surveyed regarding illnesses suffered during the preceding year. The NHS may well have been the first morbidity survey carried out by welfare-relief recipients. Around 6,000 unemployed white-collar workers were paid by the Works Progress Administration (WPA) to conduct the survey at an initial cost of $3.5 million. While it was not the first survey that verified interviews by contacting doctors, the NHS may have been the first to pay physicians for such work. Data from business and industrial health plans was also collected. This was certainly the first morbidity survey utilizing a sizable public relations apparatus. Its planning was a public event in which cities lobbied actively to be included. National media announced the start of the survey and a local media campaign was launched as soon as the survey arrived in a city. The dissemination of its results set new standards of media attention. Not least, this was the first national survey designed, in the words of the about-to-retire surgeon-general, "to study the extent and nature of disability in the general population, with special reference to *chronic disease and physical impairment*."[9] This surprising orientation grew out of the political movement to reform American healthcare that got underway in the late 1920s and intensified during the 1930s under the Roosevelt presidency.

The two men who initiated the survey—Edgar Sydenstricker and Isidore S. Falk—are familiar figures to historians of American healthcare reform. Sydenstricker was arguably the leading epidemiologist of his generation. A researcher for the PHS since 1915 (including a two-year leave to work for the League of Nations), he had participated in Joseph Goldberger's famous pellagra investigations and gone on to pioneering work in the equally celebrated Hagerstown morbidity studies of the 1920s. In 1928 he became scientific director of the Milbank Memorial Fund while remaining a consultant to the PHS. His work exemplified growing collaboration between private foundations and governmental agencies. Among his many activities was membership on the Committee on the Costs of Medical Care (CCMC), which began its work in 1927 with funding from eight private foundations. Sydenstricker was a member of a group of progressive committee members led by C.-E.A. Winslow of Yale University, who served as vice chairman of the CCMC and chairman of its executive committee (which included Haven Emerson and Michael Davis, now with the Julius Rosenwald Fund). Winslow arranged for Falk, a former student, to be appointed director of research. In this capacity, Falk directed a morbidity study of nearly nine thousand families. Enumerators visited families at two-month intervals over a full year and noted occurrence and duration of illness, as well as use and costs of medical services.[10]

Finishing its work in 1932, the CCMC failed to reach a consensus. A majority report recommended that group practices involving doctors and other health professionals become the major providers of preventive and therapeutic services. It also endorsed the extension of basic public health services to the entire population through the expansion of public health departments. Most controversially, the report advocated group payment of healthcare through insurance or taxation. The principal minority report, written mainly by physicians on the commission, recommended that government healthcare be restricted to the indigent population and those suffering from diseases that could be cared for only in government institutions; government care for the poor should be expanded, "relieving the medical profession of this burden." It also recommended the development of care plans by medical societies. Yet another minority report advocated compulsory health insurance. In none of these reports was there any mention of chronic disease as a serious problem.[11]

By the time the committee published its conclusions, the Great Depression was underway and, soon after, Franklin D. Roosevelt was elected president of the United States. He brought with him to Washington several veterans of the New York public health and welfare scenes, notably Harry Hopkins as federal relief administrator and, several years later, Thomas Parran as surgeon-general. Hopkins had worked for the New York Association for Improving the Condition of the Poor, closely associated with the Milbank Memorial Trust, and had been executive director of the New York Tuberculosis Association before becoming head of New York State's emergency relief agency; Parran had been director of the New York State Department of Health under Governor Roosevelt and represented the expansionist wing of the public health movement. Figures who had participated in the CCMC and who became influential in shaping New Deal reform efforts included Edgar Sydenstricker, Michael Davis, and Isidore Falk.

Soon after the reports of the committee were completed, Sydenstricker and Falk began working on a book about health insurance systems.[12] But the project was abandoned when both were named in June 1934 to the Council on Economic Security that designed the Social Security Act passed the following year. The council took an expansive approach to its mandate to "provide at once security against several of the great disturbing factors in life"[13] and included a study of health insurance for which Sydenstricker served as director and Falk as research associate. Sydenstricker's brother-in-law, George St. John Perrott (a mining chemist who had lost his job during the Depression and gone to work for Sydenstricker), was listed as a consultant for this study. The council as a whole included other individuals who would play a key role in the NHS. Harry Hopkins was among the five cabinet-level members who signed the letter submitting the council's report to the president while Josephine Roche, assistant secretary of the treasury in charge of public health, was a member of the Technical Board. Michael Davis (another veteran of the CCMC) was on the Hospital Advisory Board.[14] Sydenstricker and Falk wrote a report calling for provision of a health program, including health insurance. This report said little about chronic disease beyond mentioning "the need for more extended study of deficiencies in many communities in the supply of hospitals, institutions for the chronic sick and of other necessary facilities." It also noted the need for further study of measures to deal with permanent disability. In commission discussions, Thomas Parran argued for the provision of public medical services for "certain groups with small incomes" on the grounds that existing health insurance funds did not meet the costs of "catastrophic" illness requiring major surgery or "disabling chronic diseases," including "cancer, syphilis, tuberculosis, arthritis."[15]

The report on health insurance was not included in the council's final report to the president and health insurance was not incorporated into the Social Security Act. Sydenstricker explained this decision as a consequence of the need to allow the medical profession to fully study the matter and to consider other solutions. But he himself did not hide his own disappointment.[16] As a consolation prize perhaps, the final report emphasized the importance of public health prevention as a method of coping with the risk of illness. Without directly addressing the issue, the Social Security Act proved of lasting significance for the management of chronic illness and disability by doing two things.[17] First, the act provided financial benefits for certain classes of people: elderly individuals received federal old-age benefits and states received grants for old-age assistance, unemployment compensation, aid to dependent children, and aid to the blind. Such financial assistance allowed some individuals who had mild chronic conditions to continue to live at home rather than in public or charitable institutions. The funds could also be used to pay for care in private nursing homes (though not public institutions), contributing to the expansion of the nursing home industry and the closing down of many almshouses. Grants to states for maternal and child welfare authorized medical care for crippled children and vocational rehabilitation. Second, the act provided the PHS with $8 million annually to allocate to state public health agencies and another $2 million for research. As well, $7 million was allocated to the Children's Bureau to advance maternal and child health and care for crippled children.[18]

Contemporaries like Thomas Parran attributed the public health provisions in the act to Sydenstricker's initiative.[19] The increase in research funding, however, appears to have been orchestrated by L. R. Thompson, director of the NIH, who in 1934 arranged for the President's Science Advisory Council to undertake a study of medical research in the PHS and then had the responsible subcommittee packed with PHS supporters like Thomas Parran, Milton Rosenau, and Simon Flexner. The subcommittee recommended increased research in cancer, heart disease, tuberculosis, malaria, venereal disease, and dental problems and suggested that "funds for the scientific work of the Public Health Service . . . be increased by the sum of $2,500,000."[20] Thomson and his colleagues, moreover, organized a letter-writing campaign to gain public and political support for this measure.

Nonetheless, the agencies involved do not appear to have been prepared for "the rather sudden and rapid expansion of public health work contemplated under the provisions of the Social Security Act."[21] Sydenstricker reported to Falk: "the Public Health Service and the Children's Bureau are frantically making plans to spend their appropriation."[22] While Congress did not approve the full appropriation in the years preceding the war, the new funds solidified the commitment of the public health leadership to an expanded vision of public health that erased the line between prevention and care. In the words of the assistant surgeon-general C. E. Waller, it was not the amount voted, clearly inadequate in relation to need, that was significant but rather the principle. "For the first time in the history of the public health movement in this country, the Congress has made a declaration of permanent policy under which it assumes in part responsibility for protection of the health of the individual within the state and has made provision for participation of the federal government in the establishment and maintenance of administrative health services for this purpose."[23]

The Social Security Act did, of course, leave one critical issue unresolved—the problem of medical care for poor and low-income families. Waller predicted "that the most important activity of the health department of the future will lie in the medical care field, at least in assumption of responsibility for meeting the problem and of leadership in working out the solution."[24] How much of this solution should depend on practicing physicians and how much on health departments was a matter for discussion but the principle was clear. Consequently, the Roosevelt administration seemed on the cusp of introducing more radical healthcare reforms. In 1936 Sydenstricker and Falk found themselves collaborating on another governmental commission, the Interdepartmental Committee to Coordinate Health and Welfare chaired by Josephine Roche that had been created the day after the signing of the Social Security Bill. It brought together leading figures from the public health and welfare domains to coordinate the work of the various governmental agencies and "to study and make recommendations concerning specific aspects of the health and welfare activities of the Government."[25] The overextended and recently remarried Sydenstricker turned down an offer to become executive director and died later that year. Fifteen technical committees would eventually be formed to deal with such issues as crippled children services, industrial hygiene, and public health nursing. A committee to examine the issue of healthcare, the Technical Committee on Medical Care, was not created until 1937. But few doubted that healthcare reform was on the horizon.

Organizing the Survey

Several months before the formation of the Interdepartmental Committee, the federal government announced a massive new program of job creation, the WPA. Sydenstricker and Falk quickly submitted a proposal for a national health survey carried out by WPA workers. Falk later claimed that he had played a key role in convincing Harry Hopkins, who supervised the Federal Emergency Relief Administration (FERA), the Civil Works Administration (CWA), as well as the WPA, that such a survey would be a good way of spending money.[26] While the claim is not implausible, after a career in medical and welfare administration, Hopkins probably needed little convincing; he also had little trouble committing by 1936 about $29 million of WPA money for 650 studies in a wide variety of domains.[27] Falk, however, did play a key role in developing the survey. Sydenstricker produced a first draft and Falk then wrote another more elaborate version after consultation with Michael Davis and Joseph Mountin.[28] Although the organization of the survey fell to others, Falk compiled a number of preliminary tables of results and presented a plan of analysis.[29] He and Perrott also went to see Surgeon-General Cummings about the survey; Perrott had no doubt that Falk viewed it as part of a wider strategy to promote national health insurance.[30] Falk suggested that he stepped out of the picture because he or others believed that his active participation would provoke the opposition of the American Medical Association (AMA), for whose leaders he embodied socialized medicine. But there may also have been some turf warfare over control of the survey between the PHS and the Social Security Board for which Falk now worked.[31] The former prevailed and management of the survey fell to the PHS and Perrott and Selwyn Collins, both young veterans of Sydenstricker's earlier Depression studies. The original plan for a National Health *Inventory* was extremely ambitious.

> First, a house-to-house canvass in ninety-five communities, located in nineteen states, representing the various geographic divisions of the country; second, an inventory of public health and medical facilities throughout the nation; third, a study of morbidity and mortality according to occupation based upon the records of sick-benefit associations in industry; and fourth, communication with every physician attending a case of illness reported in the house-to-house canvass for the purpose of obtaining his technical knowledge of the nature of the disabling illness.[32]

While considerable data was collected and published, much of it remained un-analyzed by the time the study was shut down in 1941. It was the first of the above tasks, the morbidity study, which became the central core of the NHS's contribution to American health policy.

Once money was allocated, the survey was organized with remarkable speed; it was up and running within five months of grant approval. WPA workers with interviewing and computational skills were transferred from other projects and the survey was completed in four months.[33] By the time it was over, nearly 2.7 million urban dwellers in 84 cities and close to 150,000 individuals in rural areas had been surveyed. It took somewhat longer to collate, code, and analyze the data. The process was slowed by a strike by workers in 1937 protesting a decision to lay off women eligible for Dependent Children's' Aid (the order was rescinded) as well as other labor problems.[34] When analysis of data was still delayed, Perrott was forced to move to Detroit and assume technical as well as administrative charge of the survey. Parran ordered Dr. Lewis Thomson, head of research at PHS and the survey's titular head, to get PHS personnel working on the survey "even if a temporary curtailment of the regular work is required."[35]

Cities lobbied actively to be included in the survey. Not surprisingly, New York City was especially active, with both Ernest Boas and S. S. Goldwater sending telegrams to Thomson explaining why including New York City was imperative. Clark Tibbets of the PHS was responsible for an elaborate public relations strategy that included national radio broadcasts and print articles in local and national journals. The start of the survey in a city was preceded by a newspaper campaign aimed at promoting local cooperation. Regional supervisors were encouraged to collect and send newspaper clippings to Washington. An analyst reported on media reactions. An internal newsletter, *Progress of the Health Inventory*, was circulated to the field staff. Even before the survey was completed, results were leaked by influential public figures. The publicity barrage intensified after the report's publication.[36]

There was no official link between the NHS and the Interdepartmental Committee but Josephine Roche was responsible for both and Perrott moved from the NHS to the Interdepartmental Committee in 1937. The results of the NHS served the needs of this reform-oriented committee whose argument for health insurance and health reform generally was based on three practical arguments. (1) The health situation of Americans was very bad in the wake of the Great Depression. (2) The poor suffered disproportionately from illness. (3) The poor had much less access to healthcare than did the more affluent. As a consequence of inadequate healthcare, more people than necessary were sicker longer than necessary at enormous social and economic cost.

The problem with this set of arguments was that mortality rates had actually been falling for decades and continued to fall during the Depression. This suggested to some that no health crisis existed.[37] To counteract this perception, reformers had to demonstrate that (1) national health status could be best understood through morbidity rather than mortality rates; (2) morbidity surveys presented a somber national health picture; and (3) survey data showed that the poor, lacking adequate healthcare, also suffered disproportionately from illnesses that could be cured or mitigated. With this set of suppositions, it was possible to make a moral, economic, and medical argument for health reform, including some form of insurance or publicly organized healthcare for the less affluent. But morbidity studies were not without difficulties. Opponents could argue that surveys of this sort were highly inaccurate and that many of the illnesses reported were trivial. More problematically, such surveys had so far provided only weak evidence for the link between poverty and illness. The work of Sydenstricker and his associates, supplemented by unemployment and welfare surveys occurring simultaneously, overcame these problems by gradually focusing on chronic illnesses and longer-term disabilities.

Morbidity Studies and Chronic Disease

Sydenstricker was during the 1920s and early 1930s closely connected with the international health movement centered in the League of Nations Health Organization that focused on social causes of disease, including housing, nutrition, and income inequalities. For much of that time he showed little interest in chronic diseases, focusing instead on traditional problems, like contagious diseases and infant mortality whose incidence had declined and which held promise of further reduction and even elimination. Furthermore, Sydenstricker, like other public health figures, was dissatisfied with general mortality statistics as an indicator of national health status and guideline for healthcare planning. Similar considerations had prompted John Shaw Billings at the end of the nineteenth century to include the collection of morbidity data in the census.

Sydenstricker experimented with different kinds of data: industrial disability statistics, results of insurance medical examinations, and, especially, the morbidity surveys for which he became famous. His overriding interest was not in *kinds* of diseases but rather in distinguishing illness incidence among various groups by age, sex, ethnicity, and, increasingly, occupation and income. The most significant finding of his and other morbidity studies was that morbidity patterns differed from mortality patterns. In the case of mortality, "general diseases" like cancer and cardiovascular disease were taking a leading role, whereas respiratory and infectious diseases predominated in the case of morbidity.[38] Sydenstricker's interest in morbidity likely explains why he was not among those who warned about a "chronic disease" problem in the 1920s and early 1930s.

Sydenstricker was equally convinced that disease was closely correlated with poverty. He could draw on numerous mortality studies demonstrating this point.[39] But mortality rates were a double-edged sword that also suggested that things were improving. It proved harder to make this link through morbidity studies. Sydenstricker could point to several industrial disability studies as well as his own early work among South Carolina cotton-mill workers supporting this view.[40] But his Hagerstown studies found a poverty-morbidity link that was not strong. A detailed analysis of the data "revealed the facts that the association of illness with poor economic status 1) appeared for certain causes only, and 2) was indicated in adult life and not in childhood or adolescence."[41] On several occasions Sydenstricker speculated on the reasons for the apparent weakness of this link,[42] but he never doubted that his data seriously underestimated this correlation and that poverty strongly affected health by shaping direct causal factors like nutrition, sanitary conditions, and overcrowding. He would spend much of the early 1930s trying to strengthen the case for the relationship between poverty and disease.

The report by the CCMC convincingly demonstrated that great differences in unmet need for medical care were linked to economic status.[43] Much to everyone's surprise, however, it suggested that the more affluent suffered slightly *more* illness than did the poor. Neither the authors of the study nor Sydenstricker believed these results. They explained them away by suggesting that investigators tended to "unconsciously" record as illnesses those that entailed medical costs and ignored those that did not.[44] Consequently, the data was largely ignored. Several years later, Sydenstricker's collaborator Selwyn Collins reanalyzed the CCMC data without even mentioning social differences in morbidity rates.[45]

Most of the data for the CCMC studies was collected in 1926. The Depression created new need and opportunities for morbidity studies. In 1933 Sydenstricker began directing a survey on behalf of the Milbank Fund and PHS that was connected to an international study by the League of Nations. This consisted of a sickness and mortality survey canvassing about 12,000 wage-earning families in 10 localities, including 8 large cities, a group of coal-mining communities in West

Virginia, and a group of cotton-mill villages in South Carolina. The families were not meant to be representative; mainly poor districts (although not slums) were canvassed because the goal was to gauge the effect of the Depression on families that had been self-supporting before the Depression. Wealthy neighborhoods were excluded on the assumption that living standards of inhabitants had not dropped enough to affect health. "Colored" neighborhoods were not included to avoid the effect of race.[46]

The survey had an explicitly political rationale. Death rates and reports of communicable disease had not, it was admitted, risen during the worst years of the Depression. "The comfortable conclusion is drawn by many that the physical well-being of the American people not only has not suffered but, in view of the continued low death rate, may have been benefited by the economic catastrophe. Such a conclusion, based upon mortality statistics alone, is open to question."[47] Morbidity studies, it was claimed, constituted a more reliable indicator of health status and medical need. Sydenstricker's young collaborators Perrott and Collins reported that, with the exception of only a few localities, "the disabling illness rate of families having no employed workers is consistently higher in each city than that of families having part-time or full-time workers."[48] A major innovation was counting "disabling" illnesses, conditions that prevented people from working or otherwise functioning normally for some period; this was a standard category for industrial economists but had not been used in morbidity studies. This choice gave precedence to more serious conditions and may also have been an attempt to avoid repeating the results of the CCMC survey based on all reported illness. The excess of disabling disease rates among families classified as poor was 23% higher than in the grouping classified as comfortable by per capita income and 30% higher by total family income. Illnesses "largely chronic" that began prior to the Depression showed even higher rates of excess among the poor than among the comfortable—50% by per capita income and 80% by total family income. These numbers, which might have directed attention to chronic illness, were not followed up. Emphasized instead was the finding that families on public or private relief experienced more illness than any other group and this was true of children as well as adults. And of these, the highest rates were among wage earners in 1929 who later went on relief.[49] The authors concluded "that the highest illness rates were observed among those who had suffered the greatest change in standard of living."[50] Aside from showing greater need for healthcare by the poor with least access to it, the overrepresentation of the relief population among

ill Americans suggested that it might make good economic sense to increase availability of medical care and even to find ways to raise living standards to more
healthful levels in order to reduce welfare spending.

In September 1935, Josephine Roche, assistant secretary of the treasury in
charge of public health, made direct political use of this survey. In a *New York
Times* article, she insisted that declining mortality rates were very poor indicators
of the nation's physical condition and that illness rates were a far better gauge.
These demonstrated that the Depression had a serious effect on the "rate of acute
and chronic diseases and serious physical impairments among families on relief
rolls." She specifically cited Sydenstricker's study as having found "distressing
conditions."[51] After presenting a few statistics and emphasizing that sickness
among the "new poor" was most prevalent, she concluded in language that Irving Fisher had popularized. "Obviously facts such as these reveal not only conditions of human suffering and wretchedness but economic waste, and challenge
us to a swift-moving program of conservation of one of our most valued national
resources—the health and vitality of our people."[52]

Chronic diseases played almost no role in these studies despite the fact that
they were already being framed as a major health and welfare problem while cancer was successfully spawning an impressive institutional infrastructure. Much
of this growing concern was still largely justified by statistics that showed rising
mortality rates for cancers, cardiovascular diseases, and other chronic conditions.
But the data was highly controversial, as we have seen. A Canadian physician
writing in 1937 argued that the rise in cancer incidence was real but conceded
that the majority of commentators on the subject did not believe this to be the
case. Louis Dublin was cited as one of the few specialists who accepted the minority view; in actual fact, Dublin's monumental study of mortality written with
Alfred Lotka in 1937 downplayed the apparently huge increases in cancer rates
and even more in coronary disease rates as the result of better diagnosis and
changing disease concepts, though he admitted that as mortality from other diseases declined the proportion of deaths from these two categories had certainly
increased somewhat.[53] It is likely that sophisticated epidemiologists like Sydenstricker and Falk would not have taken rising mortality rates for these diseases
very seriously, quite apart from the fact that the overall decline in these rates argued against the need for reform. Consequently, neither had until then devoted
much time to chronic disease or invalidity. Sydenstricker mentioned it only in
passing in his synthetic book of 1933, *Health and Environment*, and the unpublished report for the Committee on Economic Security that he co-authored dis-

cussed it only briefly and tangentially.[54] But shortly thereafter, chronicity and disability assumed new significance for Sydenstricker and his team.

While Sydenstricker's group was doing its Depression study, national welfare institutions were collecting data for their own purposes and these were available to Sydenstricker's team at the PHS. In 1936 Perrott and Griffin published a paper based on a survey undertaken in 1934 by the FERA—also under the authority of Harry Hopkins—of occupational characteristics of more than 165,000 relief families in 79 cities. This was in some ways similar to earlier local welfare surveys but done on a significantly greater scale. This survey noted "physical or mental handicaps of a serious and permanent nature that impeded ability to work."[55] These could include conditions like infantile paralysis, loss of limb, mental defect or nervous condition, or diseases like tuberculosis, heart disease, or epilepsy. Twenty-one percent of those over 16 years of age reported some handicap, similar to results of another study based on medical examinations in Chicago.[56] Most serious by far were "orthopedic" problems (37 %) followed by heart and circulatory issues (33.2%), rheumatism (20.2%), and senility (20.9%). Here was an old welfare concern, chronic disease and disabilities among the relief population, now recorded by a national welfare agency and analyzed by public health experts in morbidity studies.

Sydenstricker reacted immediately to this report by shifting gears. He wrote an article published posthumously based on a section of the surgeon-general's annual report that had been written by Selwyn Collins. Sydenstricker's paper brought to the fore what had been a minor point in the 1935 paper by Perrott and Collins: the relief population not only suffers from higher rates of illness, it also "contains a disproportionately large number of persons who have chronic diseases or physical defects or who are susceptible to frequent attacks of acute illness."[57] He then went on to cite the data on physical impairment revealed by the FERA survey. "These data indicated that (a) a much higher proportion of persons on relief had serious physical defects or chronic diseases than those of the same occupational class who were not on relief; (b) in both the relief and nonrelief populations the proportion with impairments and diseases increases regularly from the lowest rate in professional, proprietary, and clerical classes to the highest among unskilled laborers."[58] Point a reinforced the link between disability/chronic disease on one hand and welfare spending on the other. Point b introduced a relatively new notion: chronic diseases were not just linked to welfare; they were far more prevalent among the poor than among the affluent, with increases directly proportional to income levels.

In a paper published that same year, Perrott made another intellectual leap by introducing "days of disability."[59] Determining the number of days of disabling illness in surveys was hardly new. Studies of industrial disability characteristically included such information while the issue of disability insurance was in the political air. Sydenstricker himself, a labor economist by training, used such data in his early studies with Goldberger on South Carolina cotton-mill workers. In fact, the working definition of illness for these studies was inability to work. Such data, moreover, was used to confirm the link between poverty and illness.[60] In his Hagerstown studies, however, Sydenstricker chose another measure of illness. *"The measure of the incidence of any specific disease was the extent to which it manifested itself in visible illness."*[61] This presumably did away with the ambiguity involved in interpreting why someone stayed away from work and was applicable to non-industrial populations. Still, information about the duration of disease was collected and it was specified that 60% of the illnesses recorded lasted eight days or longer. Sydenstricker promised in a footnote that a future paper would thoroughly analyze duration.[62] Somewhat inexplicably, no paper on this subject ever appeared and neither the Hagerstown studies nor his later Depression studies dealt with disease duration. One might plausibly surmise that such information was not considered reliable in retrospective accounts to surveyors.[63]

The CCMC survey also collected such data, which the report's authors did not bother to discuss. In his 1936 article, however, Perrott reanalyzed the old CCMC morbidity data. Viewed by annual case rates, respiratory diseases including tuberculosis predominated, with degenerative diseases not very significant. But when Perrott measured sickness disability "expressed in terms of total duration, time lost through disabling illness, and days of confinement in bed"[64] everything changed. Chronic disease was transformed into a central cause of morbidity as well as mortality. "The average total duration, the average amount of disability, and the bed days per case for this group, represented by the degenerative diseases, rheumatism, and nervous conditions, are of a definitely higher order of magnitude than those for the typically acute illnesses caused by the minor respiratory and communicable disease . . . illness due to chronic disease, although relatively low in incidence, becomes of major importance when the severity of the average case is considered."[65] Perrott closed the circle by returning to the question of illness and economic status. Although frequency of illness in the CCMC study was more or less the same or slightly greater for the well-off than for the poor, things looked very different if one focused on duration of disability. This showed, according to Perrott, that the poor suffered far more days of disability than the rich. Those earning under $1,200 annually had more than twice the disability days of those earning $3,000 or more.[66] All the pieces were now in place. By quantifying chronic diseases and disabilities as days of disability, one could demonstrate how pervasive these were and how much more the poor, with less access to healthcare, suffered from disease. Elementary social justice and economic self-interest should dictate a reform of healthcare to correct this situation. In this way, chronic disease and disability became the focus of the National Health Survey.

A Survey of Chronic Disease and Disability

Though it was shut down before most of its data was analyzed, the National Health Survey produced an enormous amount of information. In the twenty years that followed, more than two hundred reports, articles, and comparative studies based on this survey were published.[67] But the immediate most widely reported results of the study focused on the morbidity situation in the United States, purportedly serious enough to demand major reform of the healthcare system and the expansion of public health institutions. Well before the survey's completion, Josephine Roche described its expected consequences:

> The survey provides national recognition of the fact that the health service of the future will probably be expanded to cover other fields than control and prevention of the communicable diseases. With the cooperation of the medical profession, the control, prevention, and cure of all the ills of the flesh must be the ultimate goal of the health department . . . It is becoming widely recognized that physicians and hospitals cannot be expected to render service to the indigent without remuneration and that there must be public responsibility for the medical care of these unfortunates who otherwise must depend upon the charity of physicians.[68]

The survey did indeed make the point that health conditions were bad with the help of some sleight of hand. The units measured by the survey were "Disabling illness which had kept persons away from work for seven consecutive days or longer during the 12 months preceding the day of the canvass; and other handicapping disease or condition including orthopedic impairment, blindness and deafness."[69] Using such broad and largely unprecedented criteria, it was not hard to provided sensational figures. The *New York Times* blared on its front page that 6 million people daily were incapacitated in the United States. The survey's preliminary report framed it differently:

it is estimated that 23,000,000 persons, or more than one person in six in the United States have some chronic disease, orthopedic impairment or serious defect of hearing or vision. By reason of these disorders almost a billion days annually are lost from work or other usual pursuits and a minimum of 1,500,000 persons are disabled for such long periods of time (12 months or more) that they can be considered permanent invalids.[70]

Time Magazine reported on the survey in its inimitable style by connecting it to the construction of the PHS's new research center in Bethesda.

Surgeon General Thomas Parran was unduly grave as he broke ground for a new group of U.S. Public Health Service research buildings near Washington a fortnight ago. No one in the nation knew better than he the necessity of hurrying the construction and use of the establishment. For in his Washington office lay a heap of data, accumulated by inquiring WPA workers, showing how sick the people of the U. S. actually are.[71]

A closer look at the reports makes it clear that true prolonged illness was defined as three months or more of disability and that about 45% of these classified as ill in the study actually suffered from chronic disease or disability defined in this way; only 7% were deemed permanently disabled, defined as inability to work or resume normal activities for a year or more. But the one-in-six figure (or in some sources one-in-five) nonetheless stuck and would be cited for the next twenty years as the incidence rate of chronic disease and disability.

There were, of course, real causes for concern. About 56% of the permanently disabled were from 25 to 65 years of age. Aside from the general health crisis ostensibly revealed by the survey, analyses by income categories developed for the Depression studies concluded that frequency of disabling illnesses and resulting days of disability were far greater among the lowest-income group, and especially great for those on relief. The fact that low-paid workers had higher death rates than professionals was known and had recently been confirmed by an independent study.[72] But the NHS now could demonstrate health inequalities in terms of days of disability. In relief families, disability per capita was 8 days, in the poorest non-relief families 6 days, while above the $1,000 income level the figure was 2 to 3 days. Chronic illnesses of less than 12 months' duration caused twice as many days of disability per capita among relief families as among families above the $1,000 annual income level. Sixty-one percent of all permanently disabled cases occurred in the relief and lowest-income classes.[73] Black Americans had the worst disability rate of all—43% higher than that of whites—"due chiefly to the chronic diseases." But their health status improved, as with whites, along with rising income. "Low economic status, rather than inherent racial characteristics in the reaction to disease thus appears to account in large measure for the higher disability rate observed among Negroes."[74] A later study showed that low-income groups also received considerably less medical care than did higher-income groups.[75] This confirmed the earlier results of the CCMC report and reinforced the case for some form of public healthcare for the poor.

The survey produced one other conclusion critical for the purposes of this book: poor health was due essentially to chronic diseases. "The chronic diseases are the major causes of permanent disability . . . it was found that 94 per cent of all institutional cases and 74 per cent of all noninstitutional cases were included in the groups comprising the major chronic diseases and orthopedic impairments."[76] Among institutionalized cases, mental and nervous diseases made up by far the largest category followed by tuberculosis. Among the non-institutionalized, orthopedic impairments and degenerative diseases (a smorgasbord category including cancer, diabetes, cerebral hemorrhage and other forms of paralysis, diseases of the heart, arteriosclerosis, and high blood pressure, as well as other diseases of the circulatory system) were the leading causes of disability. The conclusion suggested by such findings was clear. "The control of the chronic diseases is thus an important step toward the reduction of the economic effects of disability."[77] Among these economic effects were high health costs. Chronic conditions accounted for a disproportionate quantity of medical care. The average chronic orthopedic case received seventeen physician consultations. Tuberculosis and mental patients absorbed a major proportion of American hospital facilities because of their long residences. The vast majority of patients in these two categories were in institutions operated by state and local governments so that taxpayers footed the bill for frequently inadequate care.[78]

There was one other finding that would prove significant for the future development of the chronic disease movement. Supporters of better chronic care had usually underplayed the link between chronic disease and old age, which implied medical hopelessness. But Perrott did not hesitate to emphasize that "In old age, the chronic diseases reach their highest frequency, and account for the major volume of disability . . . While persons over 65 years of age represented only 6 per cent of the population, one-third of all illnesses due to the degenerative diseases occurred in this period." The conclusion was that "the needs of this period for medical care place a special burden on old persons of limited income."[79] It also placed a special burden on health planners, who needed to estimate the effects of the predictable aging of the population.[80]

The results of the NHS were widely publicized even before the initial reports were completed. Perrott himself was tireless in publishing the results in every possible venue, including *JAMA*, the journal of the AMA. That organization had from the beginning not been enthusiastic about the survey. When its secretary, Owen West, received an invitation from the surgeon-general to send representatives to Washington to help plan the survey, West responded in a barely civil tone. "I am positively convinced that a great many physicians in various states are completely "fed up" on surveys and are inclined to be very skeptical about any kind of survey that may be proposed in their communities. Their attitude, in my opinion, is due to the fact that a very considerable number of surveys have been made under the direction of persons who have no knowledge concerning the basic values of medical service, while the field work has been done by relief workers and persons whose names appear on relief lists."[81] Perrott remembered years later that when two representatives of the AMA did come to Washington and were offered a view of questionnaires and coding books, they declined, explaining they were not interested in government "red tape." Perrott did not remember any published criticism but knew from conversations that leaders did not like the survey: "the general idea was that the Health Survey was done by WPA workers leaning on shovels . . . probably the WPA workers made things up at home because they were too lazy to go out."[82]

In fact, the AMA did respond publicly. Facing the prospects of state health insurance and a government antitrust suit against the organization launched in 1939, its leaders disparaged the survey's results. Morris Fishbein, editor of *JAMA*, dismissed them on the grounds "that only experts who could qualify as diagnosticians and psychologists could make such an undertaking worth while, not relief workers."[83] He claimed as well that the United States had the lowest rates of mortality in the world for infants and infectious disease generally. Fishbein and others argued that the problem of illness depended less on medical care than it did on economic conditions over which doctors had no control.[84] Results of a 1940 NHS study indicating the poorest groups received less healthcare than the more affluent was characterized as "having little novelty." Inequalities of every sort were widespread in the United States. An editorial in *JAMA* claimed that such inequality was far smaller "for medical service than for any of the other essentials of life." Those on relief and having an annual income of less than $1,000 received physicians' care in 78% of their illnesses, while the percentage for those with an annual income of more than $5,000 was 89%.[85]

Things heated up as a result of a discussion in the *American Journal of Hygiene* by an NIH researcher regarding the work of survey enumerators. His criticisms were technical rather than political. While they had been carefully chosen and trained, enumerators varied considerably in the proportion of illnesses they reported; the problem needed to be addressed if surveys were to become more scientifically rigorous.[86] This gave some critics of the NHS the opening they needed. An unsigned note in *JAMA* probably reflected the views of many physicians when it called into question the reliability of the survey results. Worse, the note suggested that "promotion and employment depended on the number of illnesses reported" so that the organizers got the results they were looking for.[87] The author of the original article responded that his goal was to improve the quality of surveys and that his conclusions should not be taken out of context. But he did not specifically deny the charges.[88] Within the public health movement, Haven Emerson also did not disguise his feelings about the NHS and its conclusions. Sickness incidence, he argued, was not a measure of the nation's health and the PHS would do well to launch a survey of how states were doing in supplying "indispensible services of known worth and cost to its population . . . then indeed we might have a national health survey worthy of the name and commanding universal respect."[89]

The National Health Program

One reason for the vigorous responses, both positive and negative, was that the survey was being utilized to justify health reform at a variety of levels. An internal report of 1938 by the PHS's Office of Public Health Education specified that in addition to its other educational activities, the office needed to "make the maximum impact with the National Health Inventory shortly to be released, especially as it concerns the need for better public health facilities and new responsibilities in medical care." The report went on to say that discussion of the survey "should be treated with care so that it may not be regarded as direct propaganda."[90] The AMA indeed saw the survey as "direct propaganda." In 1937 the Technical Committee on Medical Care began its work. It represented most of the federal agencies dealing with health and welfare. Among its members were I. S Falk, then working for the Social Security Board, and George St. John Perrott, Joseph Mountin, and Clifford E. Waller of the PHS. This committee produced the National

Health Program that served as the subject of a National Health Conference held in 1938 and as the basis of subsequent unsuccessful healthcare legislation.

The formal reports of the NHS, in fact, began to appear in 1938, the year that the National Health Program was published. Although created at different times and for different purposes, the NHS and National Health Program were closely connected. Both were directed by Josephine Roche, now retired as assistant secretary of the treasury in charge of public health but who remained head of the Interdepartmental Committee. And while he tirelessly published reports and articles on the NHS, Perrott also served as secretary of the Technical Committee on Medical Care and as chief organizer of the conference to promote the National Health Program. The publications of the Interdepartmental Committee ceaselessly quoted the statistics produced by the NHS. And many of the articles reporting on the results of the NHS concluded with discussions of the National Health Program.

The five major recommendations of the National Health Program were as follows: (1) expansion of public health and maternal child health services under the Social Security Act; (2) federal grants to the states for hospital construction and defraying costs of the first three years; (3) federal grants to the states toward the costs of a medical care program for the needy; (4) federal grants to the states toward the costs of a general medical care program; and (5) federal action to develop a program of compensation of wage loss due to temporary and permanent disability.[91] Chronic disease was not specifically mentioned in summaries of these recommendations. Similarly, in the discussion section of the report, the lion's share of space was devoted to maternal and child health, with traditional public health concerns like tuberculosis and other infectious diseases receiving considerable attention. Nonetheless, chronic diseases were not ignored and it was affirmed that "Only a concerted attack on these diseases as recognized problems of public health importance can hope to bring any reduction in the deaths and disability due to these causes." A comprehensive cancer program was one of the most urgent priorities as was, from a financial perspective, a coordinated program to deal with mental illness. Furthermore, several of the five recommendations were directly relevant to the chronic disease program. The report specified that facilities to care for chronic cases should be included in the hospital building program.[92] Provision of medical care for the medically needy as well as wage compensation for temporary and permanent disability were significant aspects of the chronic disease reform.

One of the specialized reports prepared for the National Health Conference of

1938 by Perrott was devoted to the chronic disease problem. But the subject elicited little discussion at the conference in comparison with point 4, the general medical care program, which provoked vigorous AMA opposition. Silence, however, did not imply indifference. Chronic disease was a noncontroversial issue that aroused virtually no opposition. The National Health Program led directly to the Wagner Bill that died in committee in 1941. The Interdepartmental Commission ceased functioning in that year, following the establishment of the Federal Security Agency, responsible for the coordination of health and welfare agencies. Chronic disease, however, was now irrevocably on the national health agenda.

Conclusion

The National Health Survey had major long-term consequences. On the technical level, the combination of public health morbidity and welfare disability studies introduced, or more correctly reintroduced, a powerful new tool—"days of disability"—that quantified and standardized the amorphous notion of "chronic disease." It was flexible as well. Depending on context and motivation, one could define such disease as seven days, three months, or one year of inability to work or function normally. The NHS and the next major survey organized before World War II by the PHS and the Milbank Memorial Fund, the East Baltimore Longitudinal Study devoted to chronic disease, as well as the first National Cancer Survey begun in 1937 by the National Cancer Institute[93] arguably played a significant role in the postwar turn of American epidemiology to chronic disease.

The NHS also had major political implications. Daniel Fox has argued that NHS data played a key role in policies to plan and construct hospitals, fund biomedical research, and expand education for the health professions.[94] The result most emphasized initially was that low-income groups suffer disproportionately from diseases and disabilities and are least able to afford healthcare. This supported arguments for some form of public healthcare for the poor, arguments that led eventually to Medicaid. In the longer term, the survey established in the public consciousness that "chronic disease" was a major public health problem. In 1940 the American Hospital Association and the American Public Welfare Association published a statement about the need to improve institutional care for the chronically ill.[95] After World War II, chronic disease became a major health policy issue, buttressed by the now-dated statistics of the NHS. The shift that occurred during the interwar years is exemplified by the title of the book that Ernst Boas, arguably the most influential spokesman on behalf of the chronic disease problem, published in 1940. While his earlier book of 1929 had been modestly titled *The Challenge of Chronic Diseases,* the book he published eleven years later was called *The Unseen Plague: Chronic Disease.*[96] Chronic diseases had been transformed from a plural to a single reified entity, befitting a political issue that needed to be easily understood by the public and members of Congress. And it had been upgraded from a mere "challenge" to a formidable "plague" all the more dangerous for being "unseen." This rather florid title reflected the sense of urgency that was beginning to spread in the United States and that would intensify dramatically during the postwar decades.

CHAPTER FIVE

Mobilizing against Chronic Illness at Midcentury

The National Health Survey on Chronic Illness (NHS) catapulted the problem of chronic disease onto the national American political stage. In the decade that followed World War II, the issue became central to a variety of American institutions: public and private, local and national, healthcare and welfare. Literature on the subject dwarfed the already significant interwar writing on this theme. Many conferences, committees, and commissions were devoted to it. New departments and jobs in public institutions were created to help deal with it. Emphasizing the problem of chronic disease continued to be a strategy that some used to advocate for healthcare reform and a national system of health insurance.[1] Nonetheless, chronic disease became largely independent of the health insurance debate during these years. Some state legislatures made significant efforts to develop institutional frameworks to deal with the issue; chronic disease programs were developed in Connecticut, Illinois, and Maryland in 1943; New York State in 1944; Indiana in 1945; and California in 1947. Chicago, Milwaukee, and Philadelphia, set up central services for chronically ill people. In Syracuse, New York, a homecare demonstration was set up to deal with the recurrence of illnesses (predominantly chronic) following discharge from hospitals.[2] And such developments were merely the beginning of a process that would intensify considerably in the decade that followed. There were many reasons for this explosion of activity.

First, the prewar National Health Survey had convinced many opinion leaders that the incidence of preventable chronic illness was far too high. That one-in-six, or one-in-five, individuals suffered from a chronic disease and/or disability, however questionable statistically, metamorphosed into indisputable fact. Arguments about whether chronic disease mortality was rising became largely irrelevant. The growing ability to control infectious diseases meant that the *proportion* of deaths from noninfectious diseases was rising ineluctably, making chronic diseases, as repeated in various formulations, "the nation's number one problem." As the President's Commission on the Health of the Nation phrased it in 1952, even though much of the increase in chronic diseases could be attributed to better diagnosis and an aging population, "these changes in mortality mean that medicine and public health must shift emphasis, both in practice and in research from diseases of the young to diseases of the aging."[3] The developing field of epidemiology discovered innovative ways to measure chronic disease morbidity and eventually create new disease categories, a process to which the developing pharmaceutical industry soon contributed substantially.[4] Finally, improved ability to prevent death from many conditions also seemed to swell the ranks of the chronically ill population.

Second, the threat of chronic disease was increasingly unacceptable to a nation that had come out of World War II as the world's dominant superpower. The need to wage Cold War (and actual war in Korea) seemed to require a healthier and fitter population. In contrast to the situation in war-devastated Europe, the natural optimism of Americans about their ability to solve problems was intensified by new wealth and power. Health, moreover, was everywhere becoming increasingly considered not merely a matter of economic productivity or humanitarianism but a basic human right.[5] And Americans, less willing to bear even minor suffering and believing profoundly in the power of medical science, were demanding public action, as they became avid consumers of the products created in ever-larger quantities by the growing pharmaceutical industry.[6] Finally, reformers could point to the striking successes of the armed forces in the field of rehabilitation and present these as models for civilians previously considered hopelessly disabled. Other medical discoveries, like antibiotics, reinforced faith in the potential healing power of medicine.

Third, and perhaps most significant, efforts in almost all Western nations to bring administrative order to healthcare institutions that had been created haphazardly over centuries by public and private initiatives and that excluded large swatches of the population took a fairly unique turn in the United States. National systems of health insurance or health services that formed the basis of postwar reorganization in Europe were unacceptable to many Americans. While the battle for health insurance was never abandoned, Americans pragmatically focused on those health issues that could command general consensus. Consequently, policy centered on problems around which broad coalitions of institutions and associations could crystallize. While each of these alliances was somewhat different, the consensus achieved frequently involved large governmental investments, accompanied by general guidelines or regulations regarding their use, combined with maximum freedom for local authorities and/or experts and professionals in utilizing these funds. In this context, chronic disease had a unique status. It was a significant component of several initiatives: the expansion of public health activities, provision of adequate hospital services, financial aid for disabled and older people, and massive financial investment in biomedical research. Simultaneously, however, chronic disease was treated as an autonomous, uniquely complex problem that required multifaceted solutions. This occurred because a coalition of influential groups and organizations found it useful as a framework for healthcare reform.

Groups Come Together: The Joint Committee on Chronic Illness

A pattern of private-public planning initiatives developed in the aftermath of World War II to address healthcare issues. A particularly influential example was the Commission on Hospital Care whose creation in 1943 was spearheaded by the American Hospital Association (AHA). Rosemary Stevens astutely describes this commission: "Its membership provided a microcosm of what voluntarism meant in American society: cooperation among major foundations, service providers, industrial and university leaders, and representatives from other walks of life, who were supported by government but who worked together to develop a program that would obviate the necessity of government control of major enterprises."[7] Much the same can be said of the coalition that gradually developed around chronic disease.

Different sectors interested in chronicity began to draw together during the interwar years. Since 1936, the AHA and American Public Welfare Association (APWA) had been meeting regularly to discuss hospital care for the indigent. Both shared an interest in increasing the public funds that paid for such care. In 1940 the two organizations established a joint committee that prepared a report, *Institutional Care of the Chronically Ill.*[8] Ernst Boas and Mary C. Jarrett of the Committee on Chronic Illness of the Welfare Council of New York worked with the committee in drawing up the statement. The problem, as these organizations saw it, was that almshouses were gradually being emptied of able-bodied individuals due to the Social Security Act and other assistance programs. Almshouses were thus increasingly being used to house the chronically ill, a task for which they were in most cases wholly unsuited. Private nursing homes were also proliferating due largely to funds provided by Social Security; many of these also lacked proper facilities. The purpose of establishing guidelines for chronic care institutions was to ensure that adequate standards would be met.

In 1945 the two organizations drafted a model hospital licensing law that included provisions for the inspection and licensure of nursing homes. The APWA seems at this point to have been taking the initiative in fostering collaboration with other organizations. In 1945 its Committee on Medical Care met with two representatives of the American Public Health Association (APHA), Joseph Mountin and I. S. Falk, to discuss chronic disease. It was decided that the director of the APWA should write to Senator Claude Pepper, who chaired the Senate Subcommittee on Wartime Health and Education. This subcommittee had done much to publicize the fact that about 40% of military-age men had been declared physically or mentally unfit for military duty. In February 1945 it published an interim report arguing that returning veterans would need far more resources than the Veterans Administration had available and that more health facilities were required for veterans and the population as a whole. The director of APWA thus requested that Pepper's subcommittee conduct a series of hearings in order to produce a special report on the care of the chronically ill.[9] The initiative was not successful, but in November 1946 representatives of the Medical Care Committees of the APWA, AHA, and APHA met for the first time. The American Medical Association (AMA) was not at the initial meeting but was soon invited to join on the grounds that any solution to the problem had to involve physicians.[10] Embattled on the political front and struggling against the Truman administration's plan for national health insurance, the AMA proved to be enthusiastic about developing closer ties and common strategies with the other groups. The four organizations each appointed two representatives to hold joint conferences on "national and local planning for chronic illness." The Joint Committee on Chronic Illness produced and published in three of the four organizational journals a policy statement called *Planning for the Chronically Ill*. This brought together the priorities of the four organizations and carefully avoided points of disagreement like compulsory health insurance.

The statement began with the by-now generic proclamation that the conquest of acute communicable diseases had focused attention on chronic diseases, now the major causes of death and disability. Eight states and four cities had initiated or were planning programs to deal with such illnesses. Consequently, the problem was not lack of interest but "a great need for comprehensive planning to insure that the widespread interest in chronic disease is channeled into sound and effective activity." The four organizations thus prepared a guide for developing community programs focusing on the "health and medical aspects of the total problem, including prevention, research, treatment and rehabilitation."[11] Citing the results of the NHS, it nonetheless excluded from discussion two of the most common illnesses, tuberculosis and mental illness, for which special provisions already existed. On this point and the insistence that chronic illness was not a problem of old age, the statement followed the example of interwar activists like Ernst Boas.

The statement's opening principle was that the "basic approach to chronic disease must be preventive." Prevention meant stronger health department programs to control chronic communicable diseases like tuberculosis, syphilis, hookworm, and malaria; extensive efforts to prevent accidents in industry, on farms, and in the home; and promoting health through child and school health, nutrition, mental health, and housing programs. Periodic medical examinations of "apparently well persons" needed to integrate the newest technology. It was hoped that current interest by the medical profession in such developments as chest x-ray surveys and diagnostic centers for cancer might signal a foundational shift in adult care comparable to what had already occurred in obstetrics and pediatrics "in which preventive supervision and examination of presumably well persons is a major requirement of good medical practice."[12]

A second major theme was the need for research to advance prevention and treatment. The experience of the war had demonstrated that "we must broaden our vision and think in terms of research planned and organized on a much larger scale than any now contemplated . . . The greatest emphasis must be placed on those diseases which are the most important causes of death and disability, such as heart disease, high blood pressure, arteriosclerosis, arthritis, kidney disease, cancer, diabetes, and asthma." The research center at the Goldwater Hospital in New York was presented as a model for such a program. More original was the call for research on administrative services for the chronically ill and on social and psychological aspects of chronic illness, including "the effect of chronic illness on the individual's social relationships."[13]

Turning to medical care, the statement called for shift in priorities from institutional care for advanced stages of disease to discovery of disease in its early stages in order to prevent or delay further deterioration. Cancer control was the chief model for this strategy, which required the construction of many more hospital and laboratory facilities, additional personnel, and greater coordination of facilities. Education about early signs and symptoms was necessary as were steps to "remove the basic economic barriers to early diagnosis and therapy." Social

factors, it was stated, could not be neglected but the only specific suggestions offered were to observe cured patients for signs of disease recurrence and to provide
occupational retraining and placement for jobs compatible with patients' fragile
condition. The report went on to list some necessary characteristics of chronic
care. It should be inseparable from general medical care, reflecting the victory of
E. M. Bluestone over Ernst Boas in the prewar debate over integrated versus separate chronic facilities. All forms of care had to be coordinated since patients were
likely to move from one kind of care to another. Hospital should be open to all
economic strata. Admission of paying patients would provide institutions with a
measure of financial stability and thus guarantee quality of care. Most fundamentally, emphasis "should be placed on home and office care," with hospital and
convalescent care, as well as rehabilitation, aiming to return individuals to productive community life, while nursing homes provided for those unable to remain in their homes. Social Security measures to maintain income were "of vital
importance."[14]

Chronic care in most hospitals, it was argued, was both inadequate and unnecessarily costly. The passage of the Hill-Burton Act to finance new hospital
building and renovation (see chapter 6) provided federal aid for chronic-care facilities (up to a maximum of two beds per thousand population) and had already
stimulated new construction. But many of these new institutions were in "areas
remote from the medical center and the general hospital and with no relation to
them" and would result in medical isolation and scientific stagnation as was the
case with tuberculosis and mental hospitals.[15] Facilities for long-term illnesses
should be located "in the very closest relation to teaching centers and general
hospitals." Specialized chronic disease hospitals were acceptable in large cities
on the grounds of or close to teaching hospitals so that research could be carried
out; these could also serve as specialized consultation centers for entire regions.
But most patients should be cared for in specially designed units of general hospitals. These units were best located in a special wing designed to provide occupational, physical, and recreational therapy, as well as comprehensive rehabilitation
services.

Nursing homes should be governed under state licensing laws with minimum
standards and regular inspection and should have arrangements with nearby hospitals for medical services. Welfare agencies had to raise their payments to privately run homes in order to fully cover costs of care. Federal and state ceilings
on payments for public assistance should be liberalized or eliminated altogether.
The Social Security Act had to be amended to provide financial assistance to pa-

tients who wished to enter public institutions. (It was now limited to those in private institutions.) The widespread conversion of almshouses to public nursing homes needed to be made conditional on provision of adequate facilities and proximity to general hospitals, with planning taking place on a statewide basis. Chronic-care hospitals and general hospitals with chronic disease pavilions should consider establishing nursing home facilities on their grounds or close by.[16]

Convalescent and rehabilitation services were, according to the joint statement, the most neglected aspects of chronic care. These services shortened the period of hospitalization, reduced recurrences, and returned a proportion of sick people to active life. They also provided an opportunity to educate individuals about their limitations and retrain them for new occupations. The statement concluded by reaffirming the need for coordination and multifaceted programs. "The total problem of chronic disease is not a series of separate problems which can be solved one by one, but rather a complex of interrelated problems which require simultaneous solution. It is recommended, therefore, that coordinated and comprehensive planning be undertaken at all levels in order to achieve effective action to meet the challenge of chronic illness."[17] Thus, the joint committee was gingerly moving away from a "one-disease-at-a-time" strategy. Perhaps the chief function of "chronic disease" during the postwar era was to transform a series of independent and discrete problems into a single, complex, and multifaceted issue, requiring massive coordination on the national and local levels. In some ways it played a role analogous to that of state health insurance systems or national health services being developed in Europe. Like the latter, chronic disease provided a framework for dealing with widely varied needs and resources.

The report attracted widespread attention. The various journals in which it appeared reported numerous requests for reprints.[18] In December 1947 the Association of State and Territorial Health Officers recommended that the Public Health Service (PHS) provide consultation to state chronic disease services and that state health departments "take the initiative in bringing together representatives of state hospital associations, state medical societies, and state welfare organizations in order to achieve effective coordinated action in the prevention and care of chronic illness."[19] The secretary of the joint committee was invited to consult with the California State Department of Public Health, recently mandated by the state legislature to investigate the chronic disease problem. At its annual meeting that year, the APWA adopted a resolution urging state and local welfare organizations to work with comparable medical and hospital organizations on this issue.

The report was meant to be a beginning; recognizing that more work was needed, the committee decided to transform itself into a "national center for chronic disease planning" whenever sufficient funds could be found.[20] The political process, which continued to be dominated by the debate about health insurance, still provided the framework for discussion, but circumstances had changed dramatically; the seriousness of the chronic disease problem no longer served merely to justify healthcare reform and/or health insurance. Rather, political forums to advance reform provided the burgeoning chronic disease coalition with multiple occasions to pursue a largely independent agenda.

The Commission on Chronic Illness, 1949–55

In May 1948 the Truman administration attempted to kick-start the health insurance debate by organizing a National Health Assembly. About eight hundred delegates representing a variety of social and professional groups as well as public agencies met for four days in Washington. Delegates were divided into fourteen sections to discuss and make recommendations on specific issues. Leaders of the Joint Committee on Chronic Illness took advantage of this forum to bring their message to a wider arena. They formed a Section on Chronic Disease and Aging that based its discussions on its recent publication, *Planning for the Chronically Ill*. Among the section's recommendations was that the joint committee be transformed into a national commission and that "its membership should be expanded to include appropriate representatives from other disciplines, organizations, and interests in the field of chronic illness." Both private sources and the government should provide funding and personnel.[21]

The National Health Assembly achieved agreement on many issues, with the significant exception of compulsory health insurance. The following year, Oscar R. Ewing, federal security administrator in the Truman administration, published a report in the name of the assembly.[22] This report was for the most part based on the unanimous conclusions of the National Health Assembly with one major exception: unlike the divided assembly, his report advocated a national program of compulsory prepaid health insurance. The Truman health insurance initiative, like earlier efforts, proved a failure. But its launch probably explains the fact that the AMA took such a prominent role in organizing the Commission on Chronic Illness (CCI) that, if hardly an alternative to health insurance, focused attention in other directions. An interim committee met in November 1948 to develop a comprehensive plan for a national commission. It was headed by Dr. James Miller, a trustee of the AMA who had chaired the Section on Chronic Disease and

Aging at the National Health Assembly. Lucille M. Smith of the PHS, who had served on the earlier joint committee, acted as executive secretary. The AMA covered 70% of the costs of the interim commission while the other three founding associations covered 10% each. The AMA also provided facilities at its Chicago headquarters, as well as secretarial and clerical services.[23]

Miller became vice chairman of the CCI, which was formally established in May 1949. The chairman was Leonard W. Mayo, then vice president of Western Reserve University and chairman of the Executive Committee of the White House Conference on Children and Youth. His background was in welfare services and he would soon become director of the Association for the Aid of Crippled Children. The commission's first director was Morton L. Levin, head of the New York State Bureau of Cancer Control and later assistant commissioner of the New York State Department of Health. He had been medical director of the New York State Health Preparedness Commission that had organized a series of studies of chronic illness several years before. Levin served for a little more than a year during a leave-of-absence from his job in New York State. While heading the commission he was one of the authors of a pioneering study on the link between smoking and lung cancer published in *JAMA*. In Levin's later version of events, the article was initially rejected. But as the commission was then based at AMA headquarters, he went to the *JAMA* offices where he convinced the editor of the significance of his article as well as another more famous one by Ernst Wynder.[24] Finding a successor to Levin was difficult due to the Chicago location of headquarters. Several obvious candidates, including Lester Breslow and Milton Terris, turned down the position. Nonetheless, the advantages of being in the same city as three of the four founding associations, not to mention the technical support provided by the AMA, proved difficult to resist.[25] Levin extended his mandate slightly, stepping down from the directorship in 1951 while remaining a consultant to the commission. Assistant director Peter G. Meek served as acting head until Dean W. Roberts, who had been the APHA delegate to the joint committee, accepted the position permanently in July 1952. Like Levin, Roberts's professional roots were in state public health administration. From 1945 to 1950 Roberts had been chief of the bureau of medical services of the Maryland State Department of Health, where he developed a medical care program for indigent people and a chronic disease hospital program.[26]

The commission had roughly thirty members at any one time, including former Surgeon-General Thomas Parran; Ellen C. Potter, deputy commissioner of welfare of New Jersey; and Lucille S. Smith of PHS. In 1953 Leroy E. Burney, assistant surgeon-general, and, a year later, Lester Breslow, chief of the Bureau of Chronic Disease, California Department of Public Health, became commission members. Individuals from the labor movement (Walther Reuther), agriculture, and industry were also appointed. A large group of thirty to forty technical advisors from the welfare, public health, and hospital fields included familiar names from the interwar period like E. M. Bluestone, Ernst Boas, and Mary Jarrett as well as many younger individuals. The commission was anything but a front for the AMA; but if the latter invested so heavily in it, it was because the program being proposed corresponded to its own evolving vision of healthcare.

The AMA continued to provide the commission with facilities at its headquarters and was by far the largest nongovernmental contributor to its budget, allocating more than $111,000 during the commission's existence. The PHS contributed more than $182,000 for special surveys, while other big contributors were the National Foundation for Infantile Paralysis ($80,000), the American Cancer Society ($56,000), the National Tuberculosis Association (nearly $32,000), and the Commonwealth Fund ($30,000 for special studies). Early on it was decided that the commission should seek "a broad base of financial support, not only for financial reasons, but equally important for maintaining wide interest in furthering the Commission's activities." It was also considered vitally important to get members from the world of business.[27] It received small sums from professional associations (including the American Psychiatric Association), national voluntary organizations (including the American Heart Association), the National Health Council, major foundations (Rockefeller, New York), four large insurance companies, the Eli Lilly Pharmaceutical Company, and John Hopkins University.[28] Despite such gifts, fundraising was difficult and time-consuming even though the CCI projected a budget of only $300,000 over five years. At one point Ernst Boas wrote Morton Levin to suggest naming to the commission Mary Lasker, the well-known philanthropist and lobbyist for cancer research and someone, Boas thought, who gave money to causes that provided her with publicity. Levin responded that she had been contacted in the past with negative results.[29] In 1950 the AMA offered the commission $80,000 on condition that other organizations provide funds to reach the $300,000 goal and, more problematically, that the CCI limit membership to representatives of organizations contributing financially. Since this would have excluded twenty-seven of thirty-one current members, the matter was finessed by the executive committee, which called for further consultations.[30]

The commission relied heavily on staff lent by the four sponsoring organizations as well as the American Nurses Association, the National League for Nursing, and the American Dental Association. The PHS in 1950 assigned Dean Krueger to the commission staff to serve as statistical analyst. In 1952 the commission moved its office from AMA headquarters in Chicago to the Johns Hopkins School of Hygiene and Public Health.[31] The ostensible reason was an invitation from the Maryland State Planning Commission to use its facilities for a study of chronic disease prevalence in Baltimore; this would continue a longitudinal study in East Baltimore begun in 1938 by Johns Hopkins researchers and which had already resulted in significant publications. The move is probably what made it possible for Dean Roberts of the Maryland Public Health Department to accept the directorship of the CCI. It also resulted in the nomination to the commission of Ernest Stebbins, dean of the Johns Hopkins School of Public Health.[32]

At its opening meeting on May 20, 1949, representatives of the four sponsoring organizations outlined the concerns of their respective groups. Dr. James Miller of the AMA, who chaired the meeting, made general introductory remarks and explained the "quick and generous response" to the new organization. "Perhaps it was because the problem of the chronically ill touched so many of us and did not implicate areas in which differences of opinion might have interfered. Common interest and purpose appeared so much more important than differences of opinion." The objectives of the commission, he declared, "appear to be acceptable to public and private agencies, to church and secular groups, and to professional and lay organizations."[33] Next to speak was Ellen C. Potter on behalf of the public welfare field. She emphasized the relationship between chronic illness and dependency on the one hand, and poverty and welfare assistance on the other. Public welfare officials had concluded that without "a frontal attack" on the problem, "the cost of providing for the chronically ill could become unmanageable in the future."[34] Edward Rogers, professor at the University of California School of Public Health, speaking on behalf of the APHA provided a brief history of public health interest in chronic illness. He concluded by referring ambiguously to the tension between a private commission and necessary public action.

> Although I am sure that the need is recognized fully by the United States Public Health Service, by the Children's Bureau, and by other government agencies, especially in the Federal Security Administration—and parenthetically, this meeting might well have been called under their auspices—there is something fundamental and very valid in the independent assembly and action of this body as representa-

tive of civilian leadership. The leadership of this Commission therefore may be looked to as the precursor and spearhead of that type of guidance and support that any and all subsequent and appropriate types of official action shall require.[35]

Dr. Albert Snoke, director of a community hospital in New Haven, spoke for the AHA and expressed the new, more expansive mood of his organization. Hospitals not only shared the responsibility of providing adequate professional care but also the obligation of offering "preventive care . . . convalescent care, and reha-bilitative assistance to patients."[36] The problem, however, was lack of money and facilities, as well as lack of interest among grass-roots members, failings that the organization's leadership was seeking to overcome. And Dr. James Miller, speak-ing now on behalf of the AMA, discussed the importance of chronic illness to physicians as well as their frustration with current arrangements. "Time and again, physicians have been hampered when confronted in the care of their pa-tients with conditions over which they have no control." He described a new mood within the AMA. "It is no longer satisfying for us merely to keep people alive. Early detection and prevention are the order of the day and with early de-tection must be a prompt follow-up and the application of those measures which would prevent the acute disease from becoming chronic." Efforts to "condition" people to undergo regular examinations had so far not been very successful. "We see in this coordinated effort an opportunity to further this necessary public edu-cation." Miller concluded by noting that some had questioned the motives of the AMA in participating in the CCI. "I wish to dispel here and now such concern. There have been, so far as I know, no ulterior motives, no special desires except to see this cooperative effort succeed."[37] The general aims of the commission were summarized in the Articles of Incorporation and Bylaws.

> To modify the attitude of society that chronic illness is hopeless; to substi-
> tute for the prevailing overconcentration on the provision of institu-
> tional care, a dynamic program designed as far as possible to prevent
> chronic illness, to minimize its disabling effects, and to restore its
> victims to a socially useful and economically productive place in the
> community.
> To define the problems arising from chronic illness among all age groups,
> with full realization of its social as well as its medical aspects.
> To coordinate separate programs for specific diseases with a general program
> designed to meet more effectively the needs which are common to all
> the chronically ill.

To clarify the interrelationship of professional groups and agencies now
   working in the field.
To stimulate in every state and locality a well-rounded plan for the prevention and control of chronic disease and for the care and rehabilitation
   of the chronically ill.[38]

In support of these aims, the commission published and distributed popular
works, assembled existing data, and determined areas of further study. It partially
financed a series of bibliographies on chronic illness published by the APHA and
served as a clearinghouse for information and new developments through its
Chronic Illness News Letter, which aimed at reaching a broad audience. The first
issue of the newsletter was mailed out to a little less than seven thousand organizations and individuals; at its height the publication had a circulation of about
ten thousand. Several thousand copies of various speeches by commission execu-
tives were also sent out, as were about ten thousand copies of a commissioned
pamphlet on chronic illness by science writer Herbert Yahraes.[39] The commission sponsored two major surveys on the prevalence of chronic disease, one in
Baltimore and another in Hunterdon County, New Jersey, a rural area with a
newly created health center; these became volumes 3 and 4 of its final report.
The Hunterdon study was financed by the Commonwealth Fund and was largely
carried out by personnel of the New Jersey State Health Department. The Baltimore study was carried out with personnel of the Maryland Health Department
and the Johns Hopkins School of Public Health.

The commission also collaborated with the PHS in a survey of eleven homecare programs. Another survey with the PHS and the Maryland State Department
of Health looked at populations in that state's public and private nursing homes.
Twelve other states eventually used this survey as a model and the combined results were published in 1957. Smaller questionnaire surveys were undertaken to
determine the activities of state medical societies, health departments, and welfare agencies in organizing or planning chronic care. The commission joined with
the National Organization for Public Health Nursing in obtaining data from visiting nurse organizations on the volume and costs of their services to the chronically ill. Another survey collected information from non-federal general hospitals
with fifty beds or more on facilities for long-term patients.[40] Among its most important activities, the commission organized two major national conferences, one
in 1951 on chronic disease prevention and another in 1954 on institutional care
of long-term patients.[41] The discussions and conclusions of these two meetings

served as the basis of the commission's overall recommendations that were pub-lished as the first and second volumes of its final report. There was talk of a third meeting on rehabilitation but time and money ran out.

During its brief existence, perhaps the commission's most significant accom-plishment was to win acceptance for the general concept of chronic disease in the face of the centrifugal forces promoting single-disease-based organizations and programs. It did this not only through its own activities but also through the ac-tions of its members who participated in all of the major commissions and meet-ings of the period. Four members of the CCI were members of the panel on chronic disease that was part of the President's Commission on the Health of the Nation organized by the Truman administration and whose final report included a section on chronic disease.[42] When rumors began to circulate that the incoming Eisenhower administration was going to establish its own commission on health and welfare, it was decided that the chairman and director of the CCI "should take such steps as seem appropriate to secure full consideration of the problem of chronic illness" by any commission set up by the new administration.[43]

At the same time as it was promoting "chronic disease" the CCI expanded the parameters of the concept. As late as 1947, mental diseases were excluded from consideration. But the commission's focus on long-term diseases transformed mental illness at least for a short time into an important element of the chronic disease problem. Similarly, the problems of older people had been traditionally underplayed among chronic disease activists. At the Health Assembly of 1948, there was heated discussion about the fact that the responsible section was called the Section on Chronic Disease and Aging and a motion to shorten the name by dropping "and Aging" was raised and defeated. In 1950 President Harry Truman held the first National Conference on Aging, and two years later the first federal funds for social service programs for elderly people were made available. All this suggested a problem that was quite distinct from chronic illness. In fact, the 1953 report of the President's Commission on the Health of the Nation discussed "chronic disease" separately from problems of aging.[44] When in that same year, the CCI's Executive Committee reported that a National Committee on Aging had been formed, it pointed out that even though three members of the CCI, includ-ing its assistant director, were members, the mandate of the new committee was: "Research and social action on all facets of the problems of an aging population, *except chronic illness.*"[45] The medical aspects of aging had now become part of the CCI's comprehensive orientation on chronic illness. In fact, little was omitted; long-term infectious diseases like tuberculosis and malaria, short-term illnesses that might lead to chronic conditions if not properly treated, and child health as a determinant of adult health were all considered part of its mandate. Proper dental care (including fluoridation of water supplies) was deemed a significant form of prevention, with dental caries designated as "an important chronic disease."[46]

The Final Report 1: Prevention

The first two volumes of the commission's final report outlined in considerable detail its comprehensive and complex policy recommendations for dealing with chronic illness. Although each treated a distinct subject—the first prevention, the second long-term care—there was considerable overlap. Both focused to a considerable degree on medical and nursing care: a major form of prevention was diagnosing and treating disease in its early stages and a key function of healthcare was educating patients in preventive behaviors. The CCI, in fact, saw them as two aspects of a multifaceted strategy implemented by a variety of groups. Consequently, the necessity for coordinating bodies at all levels was another theme common to both volumes. At the local level, the many private and public groups involved had to establish everything from planning committees to information centers; these had to be flexible and adapted to local conditions. Local groups could get advice, assistance, and money from chronic disease units established in state public health departments that ideally cooperated with similar units in state welfare departments and other related agencies. Policy should focus on what was common to different disease groups, and not be "limited to those of any special economic, racial, cultural, or other segment of the population."[47] The need for more and better-trained personnel, increased research, and innovative programs was emphasized in both volumes. As had been argued since the 1920s, it was vital to transform attitudes; chronic disease needed to be distinguished from old age; "neglect and pessimism" and assumptions of incurability had to be replaced "by an aroused social conscience and by confidence in the value of treatment and rehabilitation."

The volume on prevention was based on the conclusions of a meeting that took place in 1951. Its twenty-two recommendations, each surrounded by many pages of explanation and commentary, were based on a relatively new distinction between primary and secondary prevention. Primary prevention was about avoiding disease mainly through individual action and healthy living, making health education and promotion central priorities. Attention to nutrition, mental hygiene, rest, and exercise was fundamental. A nod was made to social factors like industrial hazards and healthy workplaces, proper housing, and personal security through minimum wages, job security, and income maintenance after retirement. But mechanisms for such policies were not discussed. The conference of 1951 identified ten conditions for which there were no certain methods of primary prevention. These included alcoholism, arteriosclerosis, and degenerative joint disease. Poliomyelitis was on the original list but had been removed by the time the final report appeared. Primary prevention, however, was possible for certain kinds of blindness, many cardiovascular diseases, cancers (particularly those associated with carcinogenic substances in industry), and dental caries. Frequently, prevention was achieved through appropriate medical treatment of conditions with chronic sequelae.[48] Doctors and dentists in private practice were to play an essential preventive role by providing appropriate treatment and health education. But government also had the critical function of supporting voluntary efforts and taking direct action in regulating environmental factors and human behaviors known to cause disease and disability (pollution control, highway safety, etc.) and investing in research. In keeping with the commission's emphasis on comprehensive action, the report recommended focusing on biological factors and causes common to many different conditions. The importance of epidemiological and social science research was also emphasized.[49]

"*Secondary prevention means halting the progression of a disease from the early unrecognized stage to a more severe one and preventing complications of sequelae of diseases.*"[50] While this could occur in various ways, the greatest potential benefit lay in "case finding," the search for early signs of disease among the seemingly healthy. The method for doing this was the periodic medical examination, now applied to dentistry as well as medicine. But case finding was now to include routine use of laboratory procedures and x-ray imaging, as well as large-scale community programs for screening, detection, and health counseling. Regular examinations were applicable to every age group but should be annual for anyone between 35 and 60 and biannual for those over 60.[51] Health examinations, it is true, were time-consuming and expensive and were not popular even when offered free of charge. The remedy was to develop simple, inexpensive procedures. Screening tests (both specific to one disease and multiphasic) conducted by technicians and then sent for diagnosis to physicians appeared to provide a relatively inexpensive mass alternative to examinations and were strongly endorsed by the commission. While screening entire populations was not feasible, screening targeted groups like hospital outpatients, students, recipients of social and welfare service, and members of insurance group plans was more practicable. And medical follow-up was critically important.[52] The committee judged that "multiple mass screening programs" represented a "major extension of the concept of preventive medicine into the chronic disease field." It was a "streamlined process assuring speed, efficiency, and economy."[53] Nonetheless, many problems had to be resolved before screening could be effective. Research was needed to construct more precise tests and determine better methods for organizing them. Their reliability and cost effectiveness had to be established. The PHS was urged to assume a leadership role in this task.

Also necessary were more health professionals trained in the "preventive aspects of chronic illness," an undertaking that required major changes in curricula. Students required exposure to comprehensive and continuous care, better awareness of emotional factors and environmental conditions influencing health status, and knowledge of epidemiology. In keeping with the holistic tone of the commission's report, it was emphasized that training for physicians and dentists required "the study of Man in his totality—recognizing that he is constantly interacting with his environment." Education for the general public was also imperative.[54] And both forms of education should focus on chronic disease generally rather than individual diseases. Morton Levin articulated this common-denominator approach.

> [P]ersons suffering from different chronic ailments may all require similar kinds of help or services, such as financial assistance, rehabilitation services, special help to enable them to continue to live at home, or care in a substitute home. They may encounter equal difficulty in obtaining or in paying for prolonged hospital care. They may also experience and need help in solving the same kind of trouble in their family relationships, the same type of emotional disturbances, or the same obstacles to obtaining employment, even though the "disease-cause" of their difficulties may vary. These are all "common denominators" from the individual's as well as the community's standpoint, in chronic disease control.[55]

Community action to provide coordinated preventive services required planning among many public and private agencies. Prevention did not necessarily require new services but that "the plethora of programs and facilities already in the community be related to each other in a rational and logical manner—in other words, that there be substituted coordinated and integrated program arrangements for the chaotic and fragmentary ones that characterize the current pattern of health services in most communities."[56] Coordination could take several forms. Screening for specific diseases by individual agencies might be replaced by collectively organized screenings for multiple conditions. Or the population might be divided up among agencies and physicians. Services in the physician's office had to be harmonized with those in community facilities. Finally, there was coordination of prevention with welfare activities and other social programs. Forms of coordination depended on local conditions. While leadership should logically emanate from the health department, it was impossible to predict which agencies or groups would assume a directing role in any given place. Whatever the local mode of organization, state health departments should provide advice and help finance preventive services.[57]

The Final Report 2: Long-Term Care

The conclusions of the commission's second major conference, revised and published in volume 2 of its final report, focused on care for "the long-term patient," a term that was carefully distinguished from chronic illness. The latter was defined as "all impairments or deviations from normal which have *one or more* of the following characteristics: are permanent; leave residual disability; are caused by nonreversible pathological alteration; require special training of the patient for rehabilitation; may be expected to require a long period of supervision, observation, or care." The former covered a much smaller group that required medical or nursing care for a prolonged period, defined as at least thirty days in a general hospital or more than three months in another institution or at home.[58] It was estimated that in 1950 about 28 million people suffered the physical or mental impairments defining chronic disease and about 5.3 million required long-term care.

Many of the eighty general recommendations made in this second volume (surrounded by nearly six hundred dense pages of text) overlapped with those of the first volume. The AMA's influence was clear in the energetic and repeated insistence on the leadership role of the patient's personal physician, which was even more energetically emphasized than in the first volume. The initial recommendation read as follows: "Care of the chronically ill is inseparable from general medical care. While it presents certain special aspects, it cannot be medically isolated without running serious dangers of deterioration of quality of care and medical stagnation." This meant that the building of special chronic disease hospitals—the pride of several cities during the 1940s and 1950s—was declared to be anachronistic. Welfare agencies continued to have a role to play in chronic care but this role was to be subordinated to that of the physician. All institutions furnishing services should reorganize "in the direction of strengthening the personal relationship of physician and patient, bringing the doctor aid and not attempting to substitute the [welfare] agency for the personal physician." The commission repeatedly reiterated the physician's central role, stating, for instance, that it was "imperative that the patient's personal physician participate as continuously as possible in the medical care of each patient at all stages of illness."[59] Another traditional recommendation was that rehabilitation, currently badly neglected, become an essential part of medical care. It was applicable "alike to persons who may become employable and to those whose only realistic hope may be a higher level of self care."[60]

A number of other themes broke decisively from the original vision of the chronic disease movement of the 1930s. Mental institutions only briefly considered in volume 1 were discussed in greater detail and were enjoined to move beyond isolation and maintenance of patients in order to become true centers of treatment.[61] The CCI's third critical principle broke from traditional interwar views that had been skeptical of the value of homecare. While high-quality institutional care remained a central goal, it was now declared that, "henceforth, communities generally should place the greater emphasis on planning for care in and around the home."[62] It was pointed out that around 80% of the disabled lived at home or in foster homes. Many more could do so under suitable conditions and in the process receive more appropriate, and cheaper, care. Making homecare central required the expansion of home nursing, housekeeping services, and technical services, measures unfortunately less popular politically than building hospitals. Successful homecare, like other services, required cooperation among many services—nursing, dental, social work, nutrition, housekeeping, and rehabilitation. The problems faced by communities were too complex to be divided up into discrete units "to be solved one by one" but required a comprehensive approach. Local communities and special state agencies would have to decide collectively what services should be offered and by whom. Communities should ideally have a single central source of counseling and referral.[63]

Hospitals remained central to chronic care but subject to the authority of personal physicians. Outpatient departments required "clarification" and reorganization in order to provide services that preserved the dignity and comfort of patients, encouraged patient responsibility for health, and "[kept] alive the doctor's interest in his patient, and the patient's respect and confidence in his physician." They should provide the diagnostic and treatment services needed by physicians for management of patients and organize homecare programs in order to provide auxiliary services to the private physician.[64] Hospitals also played an indispensable role as one among several types of institutions for long-term care. All such institutions needed standards raised "to that of care given to persons with acute illness." For hospitals to provide the specialized services needed by chronically ill patients required extensive reorganization, additional beds, and resources for rehabilitation and psychiatric care. General hospitals ideally would have chronic disease units "offering primarily skilled nursing services and physical medicine." Smaller hospitals were urged to arrange links to larger institutions in the region. "The *independent* chronic disease hospital is a second choice approach to long-term hospital care. It should be considered only when there is no practical way to associate the chronic disease facility physically and administratively with the general hospital."[65]

Another form of long-term institutional care was the nursing home. These came in bewildering shapes and sizes and many were unsatisfactory. Because they served an indispensable function, the commission recommended that the best way to raise their standards was to align them with general hospitals, through direct affiliation or working relationships. Raising standards also required increased financing through inclusion in private insurance and prepaid medical and hospital plans, action by philanthropic agencies, and increased public funding. Mental institutions were already under public control but states should prepare comprehensive plans for their improvement, with a view to enhancing "intensive treatment and rehabilitation of the patient and prompt restoration to community life."[66] For all these institutions, shortages of well-trained personnel needed to be dealt with through better education and improved salaries and work conditions. Appreciation of psychological and social factors causing and affecting long-term illness, the role of family and community, recognition of "economic, social and spiritual need," as well as team approaches to care were some of the subjects that needed to be taught. Among the first recommendations in the report (number 4) was to acknowledge the importance of the emotional state of chronic and disabled patients. "Program planning, schedules, activities, and architectural considerations must bear these points in mind." More generally, the patient needed to be treated as "an individual, whole person" who might feel "shame, inferiority, even worthlessness as a specific result of his chronic illness."[67]

In addition to classical biomedical research, national and local surveys of prevalence and incidence of disease, injury, and resulting disability needed to be carried out, along with continuing studies of healthy individuals over long periods in order to understand the "natural history" of specific conditions. Also necessary were studies to determine the economic burden of chronic disability on society and individual families. Along with the new emphasis on psychological effects and causes, attitudes about health and health services should be examined to "throw light on constitutional, emotional, cultural, and other behavioral factors that may differentiate those persons who remain relatively healthy and vigorous from those who suffer prolonged and serious disability . . . Success in rehabilitation and in the treatment of chronic illness depends . . . on the will to live, the will to do, and on spiritual values."[68] The effectiveness of treatment and rehabilitation also needed to be evaluated. All these studies required teamwork representing many disciplines and financial support from a variety of sources. A central national agency should serve as a clearinghouse for research of this nature.[69]

Financing long-term illness was recognized to be a serious problem and was discussed at some length. Both stable sources of payment for medical expenses and income maintenance were required. In the first case, the CCI took a rather complex stand, suggesting that difficult negotiations had taken place. "Until the time when the needs of all long-term patients are adequately met under plans or programs such as those listed here, there is no alternative to the basic proposition that society as a whole through taxation must meet the deficit and fill the gap. Funds for this purpose should be provided by local, state, and federal taxation. The administration should be kept as close to the person being served as is compatible with efficiency and economy. The grant-in-aid principle is applicable at both the federal and state levels."[70] The commission then hastily added: "the extension of voluntary health insurance is the primary method for financing better care of long-term illness."[71] For this to occur, however, private and nonprofit plans had to increase the number and types of conditions they covered, as well as the kinds and amounts of benefits they offered. The same applied to public programs like Workmen's Compensation. And while the CCI agreed that existing programs could eliminate many inefficient or overlapping services, it was unequivocal in demanding more generous public funding for long-term care of indigent patients in public or private institutions.[72]

The question of income maintenance created stronger disagreement and the single visible instance where a minority dissented from the majority opinion. Everyone agreed in principle that some income maintenance during illness and convalescence was necessary for proper care and rehabilitation.[73] At issue were mechanisms. Initially, it seemed as if consensus had been achieved. In the spring of 1955, the commission adopted by a majority vote a series of recommendations, including two advocating that Old Age and Survivors Insurance (OASI, or Social Security) be extended to cover income maintenance when long-term illness or disability resulted in loss of income. This in essence supported legislation then being considered by the House Ways and Means Committee. A similar bill had passed in the House in 1949 but had gone no further in the face of opposition by many groups, including the AMA. Six years later, the new version appeared with a greater emphasis on rehabilitation. The Republican minority along with the Eisenhower administration opposed the immediate passage of the bill and insisted on public hearings on the grounds that many questions about its practical consequences remained unanswered.

The conflict soon spread to the CCI. Opponents of extension, including AMA delegates, wrote a minority report repudiating the commission's original recommendation, and arguing that unlike other benefits of Social Security, disability benefits were too difficult for the federal government to administer properly. Unlike old age and unemployment benefits based on clear categorical criteria, disability was highly subjective and increased possibilities for malingering, while political pressures might militate against "sound administration of claims." They pointed to the negative experiences of insurance companies with such benefits in the 1930s and suggested these would be compounded in a federal program. They argued in addition that payment of disability insurance as a matter of right could destroy motivation for rehabilitation while raising taxes. Finally, some insisted that the law would represent "another step toward wholesale nationalization of medical care and socialization of the practice of medicine."

The opposition was successful and the commission reversed itself and formulated the following recommendation in its final report: "Further studies should be undertaken to determine whether OASI should be extended to include maintenance for persons whose loss of income is due to long-term illness or disability or whether these needs should be met other ways."[74] In a reversal of positions, thirteen dissidents, including the former Surgeon-General Thomas Parran and public health figures like Lester Breslow, prepared a minority report refuting these arguments and supporting the measure. Congress in 1956 passed the Amendments to the Social Security Act, which provided monthly disability insurance benefits to eligible workers between the ages of 50 and 65.[75]

The Commission Surveys

The Baltimore and Hunterdon morbidity studies constituted volumes 3 and 4 of the CCI's final report. These were large undertakings that significantly advanced survey methodology.[76] One of their most important innovations was to use a comprehensive questionnaire developed by the Census Bureau and to supplement interviews with clinical examinations in special clinics for a subpopulation.

(Since each examination lasted more than three hours this created considerable logistical difficulties, not to mention resistance from survey subjects.) The goal was to test the accuracy of information obtained by questionnaire and then develop a plan for error correction.[77] It was also hoped that these surveys would lead to less costly local ones by individual communities and provide greater understanding of medical and paramedical needs of the populations involved. The surveys defined chronic disease as "medically disabling" on the basis of specific characteristics, either existing or foreseeable, like permanence of change, residual functional impairment, and nonreversible pathological alterations. This definition included obesity that affected 16% of the population in Hunterdon County and 13% in Baltimore as well as dental caries. Overall, this methodology produced pathological conditions far in excess of previous studies or estimates. In spite of the fact that persons in long-term care institutions were not included in the Baltimore survey, it still found nearly 1,600 chronic conditions per 1,000 population. Prevalence rates were developed for individual diseases. In Baltimore, the rate for arthritis, usually considered the most prevalent disease, was only 75 per 1,000, whereas for heart disease of all types it was 96 per 1,000 population and for mental disorders 109 per 1,000.[78] Rates in Hunterdon County were somewhat higher (perhaps reflecting the inclusion of institutionalized patients): 120 per 1,000 for heart disease, 80 per 1,000 for arthritis, and 140 per 1,000 for mental conditions of all types.[79] Examining doctors were pessimistic about prospects for primary prevention but judged that much existing disability was not very serious and that from 30 to 40% of the cases could be improved by proper medical care (secondary prevention). Particularly alarming for the directors of the CCI was the finding that previous medical care received by 60% of those found to be ill was judged to be unsatisfactory by examining doctors.[80] This meant that much treatable illness was unnecessarily getting worse.

Like the National Health Survey, the CCI survey in Baltimore found that self-reported disabilities from chronic conditions were more common in the lowest-income groups and concluded from the data that disabling illness was more a cause than an effect of low income. The surveys were also used to compare multiple screening tests with the findings of thorough diagnostic examinations. In Baltimore these were found to be closely correlated and suggested that low rates of confirmations reported in previous screening studies might be a sign of inadequate diagnostic follow-up rather than failures of screening. The findings of the Hunterdon study in contrast were inconclusive because few subjects agreed to multiple screening tests. Both surveys, however, found that interview data was not very reliable. Only one-fourth of all conditions, including many severe cases, found in examinations had been reported in surveys, suggesting serious methodological inadequacies.[81] For reasons they could not explain the Hunterdon investigators found that the more educated the patient, the more likely the medical examination would find conditions not mentioned on the survey. (They suggested, without much conviction, that perhaps doctors took greater pains with patients of their own class and thus found more asymptomatic conditions.)[82]

In the end, the importance of the surveys did not lie in the results. By 1956, one did not need reports of high morbidity rates to believe that chronic disease was a serious problem. Their significance rather lay in methodological innovations, which identified the difficulties and deficiencies of the survey method of determining morbidity. The authors of these studies went on to play an important role in fostering discussion and debate among epidemiologists about the possibilities and limits of survey methods.[83] A decade later, follow-up studies to the Baltimore survey continued to appear.[84]

The End of the CCI

When the commission dissolved itself in 1956, it went out with a flourish. In that year, the National Health Forum, an association representing a large number of American healthcare organizations, devoted its 1956 meeting to chronic illness. Speakers included Leonard Mayo, Morton Levin, and Lester Breslow, who had all been active in the CCI and who reiterated its recommendations.[85] President Dwight Eisenhower sent a telegram citing an urgent need for research on chronic illness and the need for "prompt and widespread application of existing knowledge in this field."[86] Three volumes of the CCI's final report were published a year later (the Hunterdon study was not published until 1959) amid some fanfare but with nothing like the attention that the National Health Survey had received two decades before. The lengthy, dense, and highly complex program for dealing with chronic disease was not easy to encapsulate in newspaper headlines.

It was decided that the organizations involved should independently continue the CCI's work. The AMA's Council on Medical Service took over publication of the *Chronic Illness News Letter* and, like the other organizations, set up a committee on chronic disease. In 1955 a journal called *Chronic Disease* began publication under the co-editorship of David Seegal, director of research at the Goldwater Chronic Hospital, and Joseph Earl Moore of Johns Hopkins University who, with the advent of antibiotics, had transformed his syphilis clinic into a chronic disease clinic specializing in lupus. (As a venereologist, Moore had been one of several

Hopkins researchers involved in the studies conducted by the US Public Health Service in Guatemala from 1946 to 1948, in which several groups of people were infected with venereal diseases in order to test the effectiveness of penicillin.) In 1958 the Johns Hopkins School of Public Health (which had created the first department of epidemiology in the world) named Abraham Lilienfeld to what was probably the first chair of chronic disease in the country. (The medical school already had the Joseph Earl Moore Clinic directed by Victor McKusick and specializing in clinical research on various chronic diseases and rehabilitation.)[87] Johns Hopkins was certainly a leader in the field but by the mid-1960s it was not alone. An organization called the Conference of Chronic Disease Training Program Directors of Schools of Public Health organized meetings until about 1970, after which it seems to have dropped out of sight. It also collaborated on the production of a compendium textbook made up of seventy-five mostly reprinted papers, co-edited by Lilienfeld.[88] The new ongoing National Health Survey established by the Eisenhower administration continued the work of quantifying chronic conditions and disability.[89]

The influence of the CCI was not limited to the United States. In 1957 the European Section of the World Health Organization (WHO) held its first meeting on chronic disease. Other meetings would follow. The British read the commission's report carefully and were generally unimpressed. A. P. Thomson of Birmingham (whom we shall meet in chapter 9) declared that the "report lacks precision."[90] Still, it was not without influence in that country or other parts of Europe. Canada, America's neighbor to the north, also began to take cognizance of its chronic disease problem. Having done its own national morbidity survey in 1950 that did not discuss chronic disease at all, the medical profession was gradually told by the country's chief medical officer that survey data indicated that chronic disease was indeed a serious national problem.[91]

Holding together a notion as broad as "chronic illness" proved difficult without an organization like the CCI to lobby forcefully and incessantly for a comprehensive approach. Issues that had been lumped together with chronic disease by the CCI gradually split off. In 1955 Congress established the Joint Commission on Mental Health and Illness and from then on this issue was ordinarily dealt with independently. Other diseases continued to have a life of their own. Under the Nixon administration, the National Cancer Act of 1971 declared war on that disease. The question of the elderly population was more complex. In 1956 a Special Staff on Aging was established within the Office of the Secretary of Health, Education, and Welfare, and a Federal Council on Aging was created under President Eisenhower. In 1961 the first White House Conference on Aging was held. In 1962 legislation was introduced in Congress to establish an independent and permanent Commission on Aging. But when Medicare was added to the Social Security Act in 1965, the health problems of elderly individuals once again became identified with medical treatment of chronic disease.

Conclusion

There was not a great deal in the CCI's final report that had not already been discussed in major medical journals, in commission reports, or at previous conferences. The signal achievement of the commission was to transform a concern that had achieved significance during the 1930s as part of the drive for national health insurance into an autonomous issue of substantial import for the healthcare, public health, and social service professions, and not a few politicians. It also managed to get a motley collection of organizations and agencies to work together and think beyond individual diseases in order to face "the broad problems common to all chronic diseases. The union of diverse interests in a single cause was one of the most significant successes of the Commission."[92] Surgeon-General Leonard Scheele went further and lauded the commission as "a powerful integrating influence at the national level" that had channeled many accelerating dynamic forces "in new and constructive directions."[93] Most significantly, it provided the closest thing to a national healthcare policy that could be elaborated in the fragmented American context.

In this respect the commission report was less than perfect. Reflecting the views of a broad coalition, it was not always coherent. Despite the apparent consistency created by emphasis on prevention and rehabilitation, the reality was a collection of disparate recommendations reflecting the views of specific groups. The medical profession was guaranteed that private practice medicine would be at the center of chronic disease policy. Public health workers got their wish to be involved in healthcare through the idea of "secondary prevention" that allowed them to organize programs of screening and treatment. Those working in the chronic-care hospital and welfare sector were promised better facilities and a central role within mainstream medical practice. Leaders of the hospital sector were also promised a prominent role, with hoped-for increases in resources. Biomedical scientists and epidemiologists were encouraged by the continued emphasis on research that was already transforming the American scientific landscape. The elephant in the room, of course, was national health insurance, a subject that the CCI deliberately ignored. The recommendations of the commission thus represented a compromise on two levels: among the groups participating in the alliance; and between already-compromised principles and the harsh realities of the American healthcare system. But the watering down of initial ideals does not mean that its conclusions can be dismissed as mere rhetoric. American efforts at midcentury to develop a comprehensive chronic disease policy proved unique in the international context and led subsequently to a singular flurry of activity and investment around the chronic disease problem.

CHAPTER SIX

Long-Term Care

The Commission on Chronic Illness (CCI) set out an expansive, multifaceted
strategy to cope with chronic disease that can be faulted on many levels. Foreign commentators noted the generality of its proposals. From our own twenty-
first-century perspective, there was perhaps too much emphasis on secondary
rather than primary prevention, unrealistic expectations surrounding rehabilitation and screening, and little thought about the costs of recommended reforms
or how exactly to finance them. Mantra-like calls for "planning" and "coordination" had limited impact in a fragmented institutional system dominated by free
market values. The call for comprehensive policies bumped up against the stub-
born reality of single-disease interest groups and programs. Some types of diseases, mental illnesses in particular, were unique enough in their symptoms and
institutions to be almost automatically separated from other chronic diseases.
Cancer had its own commanding independent existence. Disease-specific programs often seemed more practical and affordable than broad, comprehensive
projects. Finally, there was the "identity" axis. Older, handicapped, and mentally
ill people each had distinct interests and concerns that did not necessarily fit
within a comprehensive chronic disease framework. For all the rhetoric about
the continuum from prevention to long-term care, the resources and policies
required to prevent disease differed substantially from those needed to care for
people who were severely and permanently infirm and disabled.

Older people were particularly difficult to classify. The majority was reason-
ably healthy and had nonmedical demands that politicians were struggling to
meet.[1] On the other hand, Medicare located them firmly within the healthcare
system and they constituted a large proportion of those requiring long-term nursing care. A series of meetings by the Committee on Aging of the American Medical Association (AMA) in 1965 on "Aging and Long-Term Care" exemplified the

dilemma. The introductory remarks emphasized that despite the meeting title, aging and long-term care were very different subjects. The committee, in fact, had gone through several metamorphoses. In the 1950s the committee tried to deal with diseases of the aged. It soon became clear that the majority of elderly people were well and emphasis shifted to healthy aging involving "employment, recreation, education, family life—even housing." By 1965, the committee was again shifting its attention to the basic problems of chronic illness and long-term care, which, it nonetheless insisted, "are not exclusive to any particular age group . . . We urge that the details of treatment for long-term patients be dictated solely by health needs and not by age."[2]

Despite its limitations, the CCI report and the forces behind it set the stage for a sustained campaign to deal with chronic illness that was unprecedented internationally. To some degree, this campaign was a consequence of the failure to agree on a national health insurance plan and served as an alternative reform strategy. Elsewhere national insurance or health services generated their own problems, tensions, and perspectives that did not make chronic illness the center of policy initiatives. But when Europeans years later also came to view chronic disease as a serious problem, they were much influenced by the vast American experience of and literature on this subject.

Daniel Fox has astutely noted that chronic illness has been for some time at the center of American policy to finance healthcare. He is certainly correct in noting that a good part of the problem is that, due to the fragmentation of insurance plans, the high costs of chronic disease and disability are spread among a much smaller proportion of the population in the United States than in other Western nations. But the situation is even more complex. Public services have also been more elaborate and expensive in the United States than in other nations because, among other features, they specifically target the most medically needy. The Social Security Act of 1935 defined those who deserved entitlements. Three overlapping groups were among those singled out: people who were elderly, indigent, or had certain disabilities. Much of the history of American healthcare has been about expanding both inclusiveness and entitlements for these groups, all requiring considerably more long-term care than other Americans. (Medicare started with older people, but the 1972 amendments extended coverage to non-elderly disabled people and sufferers from chronic renal disease.)[3] The provision of such care has thus taken up far more public space and money than in countries where publicly subsidized care is available to a wider population with more varied needs.

In this chapter I examine some of the consequences of the chronic disease movement in the two decades following the CCI's report. I begin with one of key subjects discussed in the commission's final report: the provision of proper facilities for chronic care and rehabilitation. I go on in chapter 7 to another of the CCI's priorities: the expanding role of public health. My goal here is not to evaluate whether the American response to chronic disease was successful or not. I leave such questions to others better qualified than I am to define success and failure. My goal as a historian is the more modest one of describing a national effort that was deeply flawed but also uniquely ambitious.

The Hill-Burton Act and the Expansion of Long-Term Care Resources

Expanding hospital resources for chronic illness began long before the CCI issued its report. Although the founders of the commission declared as early as 1947 that chronic illness belonged in general and not special hospitals, it took time for the word to spread. In 1950 the Jewish community of San Francisco announced the construction of a chronic disease hospital modeled after New York's emblematic Montefiore Hospital.[4] New York City was justly proud of the private Montefiore and the municipal Goldwater Hospitals and in 1952 added a second, long-planned chronic institution, the Bird S. Coler Hospital, to its Welfare (now Roosevelt) Island complex. It was almost as large as the Goldwater Hospital with three buildings covering fourteen acres and eventually holding more than one thousand beds. It merged with the Goldwater Hospital in 1996.[5] In 1952 Massachusetts began constructing a state chronic disease hospital in Franklin Park, Boston. It was to cost $13 million and hold six hundred beds (a number it never seems to have reached). It was to be state of the art, with ten operating rooms "and provision for televising operations." It was emphatically not to be a custodial, domiciliary, or terminal institution; it would specialize in diagnosis, treatment, and rehabilitation of chronic conditions with the goal of getting patients back home as quickly as possible. It was also to be a center for research and training, affiliated with the three medical schools of Boston.[6] Despite the apparent success of these hospitals, the chronic disease hospital was fading away. The most visible sign was the gradual transformation during the 1950s of the always-up-to-date Montefiore Hospital, which gradually morphed into a general hospital with chronic units; in 1963 Montefiore became the teaching hospital for Albert Einstein College of Medicine.[7]

During World War II, there emerged a general consensus that hospital facilities in the nation were inadequate. The American Hospital Association (AHA) in 1943 spearheaded the creation of a voluntary Commission on Hospital Care to study the situation and plan for postwar hospital construction. With the financial support of several foundations, the commission began its work in 1944 and produced a report in 1947. Rosemary Stevens has viewed that commission as largely a creature of the AHA, seeking ways to provide public money to hospitals without excessive governmental intervention or imposition of national health insurance.[8] The planning carried out by the commission and the state planning groups inspired by its programs had considerable influence on the implementation of the Hill-Burton Bill passed in 1945. This bill provided federal aid to states to survey existing hospitals and public health centers, plan for the construction or reconstruction of such facilities, and authorize grants to contribute to this construction, with local funds making up the rest. The Public Health Service (PHS) administered the program, providing further official confirmation of the central role of public health in curative medicine, at least in its planning phases.

By 1974, Hill-Burton had invested more than $4 billion in federal funds in about 11,550 projects, with local funds contributing a little more than $10 billion. About one-third of the projects before 1970 involved new construction and for the period 1969–74 the figure was down to about 7%; modernization of existing facilities accounted for the bulk of funding. Voluntary institutions received nearly 60% of funds (as of 1970) with public institutions receiving the remainder. During the first twenty years of the program, almost half the funded projects were in communities with a population of less than ten thousand. Southern states qualified preferentially on grounds of rural need and low per capita income.[9] The emphasis was on providing such communities with general hospitals and acute beds.

The situation began to evolve after passage of the Amendments to the Hospital Survey and Construction Act of 1954 that broadened the program to public and voluntary-nonprofit nursing homes, diagnostic or treatment centers, rehabilitation facilities, and chronic disease hospitals. A special program appropriated $21 million especially for these purposes.[10] It is likely that the publicity then being generated by the CCI played some role in this legislation. In the summer of 1956 states received even more incentive to develop long-term institutions when an amendment to the Social Security Act provided cash benefits to the disabled. The Community Health Services and Facilities Act of 1961 was designed to expand services for the chronically ill and aged populations. It doubled annual appropriations for nursing homes, raised annual research appropriations, and authorized experimental and demonstration projects.[11] Two small business construction programs in 1956 and 1959 helped stimulate the growth of nursing homes. The Kerr-

Mills Act of 1960 provided matching funds to states to create the Medical Assistance for the Aged program. This act is largely considered a failure because many states did not participate; the same cannot be said of the Social Security Amendment of 1965, which created Medicare and Medicaid and injected huge amounts of new money into healthcare that could be used for chronic and long-term care. Simultaneously, Hill-Burton money devoted to long-term care expanded significantly. Between 1947 and 1974, more than $1.7 billion out of a total of $15.5 million went toward long-term beds, accounting for 22% of projects, nearly 21% of beds, and 16% of funding. But these figures include the early years of the program devoted almost exclusively to acute-care facilities. From 1969 to 1974, the proportions devoted to long-term care were 29% of projects, 27% of beds, and 22% of funds.[12] These numbers understate the investment in long-term care since they do not include mental or tuberculosis hospitals, rehabilitation centers, outpatient departments, and public health centers. Adding these would increase the share of Hill-Burton Funds devoted to the chronic sector during this five-year period from 22 to nearly 30%.

New or modernized long-term beds were not distributed evenly. Nine states got less than six hundred beds each from the program. Three states—Ohio, Pennsylvania, and Texas—obtained more than six thousand each. A few of the largest states, like Ohio, New York, and especially Pennsylvania, emphasized nursing home rather than hospital capacity. Nationally a little more than half the new long-term beds were in hospital units, on the premise that the general hospital was to be the new center of chronic care. About 40% of beds were in nursing homes, which received an increasing share of money and attention as time went on. Only about 8% of new beds were in chronic disease hospitals, now largely seen as anachronisms. A survey by the AHA found that specialized chronic-care hospitals had lost eight thousand beds between 1960 and 1969.[13]

According to the recommendations of the CCI, the general community hospital was to be at the center of chronic and long-term care. By the late 1950s, students of management were writing graduate theses on this topic.[14] Daniel Fox suggests that most new long-term care units ended up being utilized for acute rather than chronic care.[15] This may well have occurred over the long term. But during the 1960s and early 1970s, the commitment of many hospitals to long-term care was evident. In 1966 the AHA released a booklet on how to set up extended-care hospital units and the issue was widely discussed in major hospital journals.[16] By then the hospital literature was full of calls for "comprehensive" or "total" care, as well as advice on how to achieve it. It was also rife with descriptions of new model institutions that had been or were being created. One explanation for this was Medicare, which from 1967 on provided reimbursement for extended-care services in hospitals; this proved to be a major stimulus for "complete medical care" hospitals.[17]

The hospital literature is full of articles about individual long-term units like that of Fairview Hospital, Minneapolis, which operated a hundred-bed extended-care unit. The Baltimore City Hospitals consisted of an acute-care general hospital with just under 500 beds and a chronic-care hospital with 321 beds. They were under the same administration so movement of patients between the two was relatively uncomplicated. The chronic-care hospital benefitted from resources of the general hospital, and it was claimed that the length of stay in the latter was dramatically shortened (by twenty-four days) in comparison with other acute-care hospitals in Baltimore.[18] More modestly, Chippewa County Hospital in northern Michigan had a thirty-bed long-term care unit; desiring an occupational therapist for these patients but lacking the necessary funds, it hired a part-time consultant therapist whose main task was to train a full-time assistant.[19] Other institutions were innovative in different ways. South County Hospital in Wakefield, Rhode Island, created a thirty-bed extended-care unit which it was claimed cost one-third less than conventional wards mainly because of a wide carpeted "patient corridor" that was designed to get ambulatory patients out of their rooms to interact and if necessary help each other. It also permitted family members to assist more easily in patient care.[20] A small community hospital in Greenfield, Massachusetts, fully devoted to acute care in 1961 developed a ten-year project to foster continuity of care to include extended-care rehabilitation and homecare.[21] Highland Park Hospital in Cleveland established a sheltered workshop for chronically disabled individuals.[22]

It became accepted thinking that such extended-care facilities be set up within or adjacent to the main hospital building. But this was frequently not the case. A study of long-term care units in thirty-one states conducted at Washington University, St. Louis, found that physical separation or difficult access to the main hospital was common and not conducive to high-quality patient care. Studies of long-term care units in hospitals in Wisconsin came to similar conclusions; proximity influenced transfer of patients, continuity of services, and physician participation. The ideal location for such units was in the actual hospital building. The administrator of a small hospital in Troy, Ohio, proudly declared that at his institution, the extended-care unit was an extension of the surgical corridor and operated as part of the hospital exactly as did the obstetrical and psychiatric services.[23]

A version of the 1969 Hill-Burton Amendments introduced but defeated in committee contained a clause requiring that extended-care facilities built with Hill-Burton loans and grants be physically attached to hospitals.[24] Methodist Hospital in Mitchell, South Dakota, went even further and established a retirement residence connected by a tunnel and a shared administrator to the hospital.[25] Institutions that were administratively distinct might also benefit from close proximity. The Moss Home for Incurables in Philadelphia was another private Jewish institution established in the early twentieth century. By the 1950s, it was turning into a modern chronic-care hospital and rehabilitation center. Around 1960 it moved next door to the Einstein Medical Center, a large acute-care hospital. Although the 124-bed chronic-care hospital remained administratively distinct from the neighboring thousand-bed medical center during the 1960s, coordinated activity was the rule. Specialized testing and cooking were done at Einstein, and "patients move freely in either direction between the two institutions . . . The medical chart accompanies the patient."[26]

An interesting and illustrative case of the problems caused by distance is that of Baptist Memorial Hospital, a 1,565-bed institution in Memphis, Tennessee.[27] In 1962 it took over an abandoned Veterans Administration hospital several miles away and began a $1 million renovation to create a six-floor extended-care general facility for elderly patients that became a satellite unit of the hospital with long-term care initially the major activity. It soon became clear, however, that there was little interest among hospital staff for the remote unit; long-term care proved expensive and not always covered by insurance. It was decided to transform the unit into a comprehensive rehabilitation facility for all ages thanks to a grant from the W. K. Kellogg Foundation. A lack of medical leadership was partially solved when a regional spinal cord injury center was established within it; this attracted a young orthopedist who also became part-time medical director of the rehabilitation unit. The story is instructive on a variety of levels, aside from the issue of proximity. Long-term care was abandoned not for acute care but for rehabilitation. But even run-of-the-mill rehabilitation was not necessarily attractive to highly trained physicians until a high-tech spinal cord injury center transformed the landscape. We return to the issue of rehabilitation later in this chapter.

Hospital statistics for this period are not very consistent. An AHA survey of 1967 found that 14.5% of general hospitals, mainly large institutions, had extended-care facilities.[28] Another AHA survey two years later found that 12.9% of reporting hospitals had extended-care health units, while 17% had inpatient and 14% outpatient rehabilitation units; these were mainly concentrated in larger urban institutions. In 1969, 24% of hospitals had occupational therapy departments and 59% had physical therapy departments (an increase from 41% in 1960).[29] In the late 1960s a review of the literature concluded: "The extended care facility is emerging as primarily a hospital program, with nursing homes and residential homes set apart in the area of long-term care."[30] An AHA survey in 1980 of 6,282 hospitals found 693 long-term skilled nursing units containing 44,294 beds in or attached to hospitals; 298 units without skilled nursing held 27,651 beds. Another 39 hospitals and 28 units, together holding over 24,000 beds, were listed as fully chronic disease institutions. Taken together, nearly 17% of hospitals reporting were fully or had units devoted to long-term care, not including those specializing in psychiatry, mental retardation, tuberculosis, and addiction.

Looking more closely at the 1980 statistics, one is struck by changed categories; "chronic disease units" had now completely disappeared, replaced by "skilled nursing units."[31] This was not just a change in nomenclature but reflected a real shift that took much medical treatment out of these units and replaced it with cheaper, albeit skilled, nursing care. One reason for this change is that medical practice around chronically ill patients was changing. The most serious diseases had by now become firmly integrated in hospital culture. Specialized units were now in existence to provide technological procedures or intermittent acute care; coronary care, cancer therapy, and dialysis units were gradually evolving into short-term facilities with patients being sent elsewhere as soon as intensive technology or interventions were no longer necessary. Furthermore, during the 1960s, outpatient services far outgrew inpatient services, a development which meant that many chronic patients could be treated without long hospital stays.[32] Numerous hospitals established rehabilitation units, but these were usually run independently of long-term care units. Many of the units that remained dedicated to long-term care may well have become skilled nursing units long before the change was registered in AHA statistics.

A precipitating factor for this shift was undoubtedly the rapidly rising costs of Medicare and Medicaid, which created enormous pressure to get people out of hospitals and subverted the ideal of chronic-care units. By the mid-1970s, legislation reclassified chronic disease hospitals as skilled nursing homes, viewed by many as a move to reduce both costs and the amount of available medical care. This interpretation was proved correct when Medicare officials in 1977 began refusing to reimburse elderly patients treated in chronic disease hospitals, saying they officially recognized as "acceptable for reimbursement" only treatment in general hospitals or nursing homes. Attempts by physicians and state departments of health to fight these changes had little effect.[33] Given such reimbursement policies, it made sense for hospitals to transform chronic disease units into cheaper skilled nursing units.

Nursing Homes

Whether they had long-term units or not, hospitals were being asked to become part of a system of "community" care that extended beyond their walls. Another component of this community care consisted of nursing homes. During the 1960s, nursing homes benefited directly from the building boom generated by Hill-Burton. But they constituted something of a problem. The nursing home industry was, if not created, given a major boost by the Social Security Act of 1935 that provided cash subsidies that could be used for those living in private but not public nursing homes. M. Holstein and T. R. Cole argue that this limitation originated in hatred of almshouses that resulted in resistance to public provision of nursing home care and the consequent growth of a proprietary nursing home industry. Later, direct federal payments to institutions assured reliable income and thus promoted the industry, as did construction loans to nonprofit homes through Hill-Burton and to for-profit institutions through other construction programs.[34]

A study published in 1957 by the CCI and the PHS tried to make some sense of the plethora of largely unregulated nursing homes in existence. Examining institutions in thirteen states, the study characterized the nursing home as a "new phenomenon in American life. It grew out of a complex of social, medical, and economic changes in our society. It offers to meet problems aroused by those changes. In its very emergence, however, it poses its own problems. What it offers, therefore, is not necessarily always the professionally appropriate answers to the problems it attempts to meet."[35] What made for such inappropriateness was the extreme variety of institutions. The survey created a classification to make sense of the chaotic situation. "Skilled" homes provided technical nursing skills like enemas, irrigations, catheterizations, dressings or bandages, and administration of medications. "Personal care" homes could provide help in walking, getting in and out of bed, bathing, dressing, and feeding. Finally, there were boarding establishments for individuals who could largely care for themselves. Complicating matters, many establishments had a mixed character. And unlike chronic-care hospitals, the survey found, these homes catered primarily to aged people. It was estimated that out of about twenty-five thousand homes of all types in 1954, 40% were "skilled." The majority of homes in every category were proprietary rather than public. There is "great need," it was concluded "for developing some standardization of the meaning of the designation 'nursing home.'"[36]

One strategy adopted by the American Nursing Home Association was a strong lobbying effort to sponsor legislation that imposed licensing and classification of nursing homes according to services provided. Whereas only five states had licensing laws in 1950, all states did by 1965. Licensing requirements, however, differed substantially from one state to the next. Medicare legislation, in contrast, created a new, more rigorous term, *extended-care facility*, defined as one providing skilled nursing care or rehabilitative services. These facilities were required to have one professional nurse employed full-time with skilled care available twenty-four hours a day.[37] This was a narrow medical definition that was associated with limited-term care and that only applied to homes wishing to work with Medicare. It was estimated that, at most, 40% of nursing homes during the late 1960s qualified. Thus, variation among homes continued to be wide. In Connecticut, where state payments were tied to institutional classification, it was claimed in 1967 that that the situation was improving and that many more homes were providing nursing, rehabilitation, and therapeutic services. But a survey that same year in Massachusetts found that only one-quarter of all homes provided adequate care and half had "irremediable shortcomings."[38]

Because of the large number of quite different institutions involved and the changing categories used to classify them, surveys came up with rather disparate results. A PHS survey in 1963 found a little more than 17,000 facilities of all types, including 728 nursing home units in hospitals. Another survey 2 years later located a little less than 19,000 licensed institutions with more than 760,000 beds. Of these, 12,244 with 524,000 beds were classified as full nursing homes. The number of units in hospitals had fallen to 326.[39] In 1969 it was announced that the nation's 23,000 nursing homes and related facilities (13,000 of which were licensed) had passed the 1 million-bed mark. Many homes, however, were being reclassified in order to offer less nursing service than those mandated by federal-state programs.[40] In 1977 the National Health Survey report on nursing homes found a little less than 19,000 homes of all sorts with only about 8,200 skilled nursing facilities containing slightly fewer than 800,000 beds. Throughout this period, most homes were in the private sector, either nonprofit or commercial, while public homes "languished." Along with these changes, the proportion of elderly and seriously ill patients in nursing homes increased significantly as states transferred patients from state-funded institutions like mental hospitals to nursing homes in order to take advantage of federal funding.[41]

Initially, efforts to introduce higher standards were not pursued rigorously because of fears that they would exacerbate the problems of rising costs and bed shortages. But gradually criteria for eligibility rose, with a primary focus on facilities for rehabilitation, meaning that skilled care increasingly meant access to medical care. The National Health Surveys of 1964 and 1968 monitored how many homes had doctors on staff (not many) and how many had doctors who visited regularly or who were on call.[42] Medicare's shifting criteria of acceptable care provoked considerable consternation within the nursing home industry. In 1971 the Executive Board of the American Nursing Home Association passed a resolution withdrawing official support from the Medicare extended-care program and urged its seven thousand members to "reassess their Medicare participation." The association complained that there had been a dramatic decline in the number of Medicare patients receiving extended-care benefits "due to unfair and restrictive controls from governmental authority."[43] It claimed that only 5% of Medicare-certified beds were being utilized for Medicare patients, a figure hotly contested by the Social Security Administration. Such problems persisted. It was reported that Medicare provided only 2% of nursing home revenues in 1977. In contrast, Medicaid provided half the industry's revenues and supported about 60% of nursing home residents. One result of Medicare's restrictive policies, it was charged, was that far too many patients remained in hospitals simply because there were no places for them in nursing homes.[44]

Aside from seriously inadequate medical care in many institutions, poor diet and fire were constant problems; between 1961 and 1965, 2,500 nursing home fires broke out; in 1965 alone an estimated 800 fires caused 38 deaths. Stories of patient mistreatment, spoiled food, and other abuses abounded. Conferences on the subject were held regularly and numerous professional associations weighed in on the matter; administrative bodies, including the Division of Chronic Disease of the PHS, published guidelines and statements of standards.[45] The situation seems to have improved somewhat in later decades, but other problems came to the fore. As frail and frequently demented elderly people increasingly populated nursing homes, a reaction set in to both the culture of medicalization and poor custodial care. This was most clearly expressed in Bruce Vladeck's much-cited book of 1980, *Unloving Care*, and in a 1986 report by Institute of Medicine that argued that in addition to gross abuses in some institutions, quality of life in most was poor even when medical care was good.[46] This report led to new federal regulations to monitor and improve the quality of care that did little, in the view of

Vladeck and others, to dispel the "chronic malaise" surrounding the mission and culture of nursing homes.[47]

Nursing homes remained a central healthcare issue in the United States. Between 1973 and 2004, seven surveys of these institutions were carried out by the Centers for Disease Control (CDC). And the issue periodically attracted significant attention from other institutions and the media. Increasing costs aggravated the quality problem. Although cheaper than hospitals, nursing homes were nonetheless expensive; costs increased from 1950 to 1981 by an annual rate of 17%.[48] Worst of all, they did not solve the problem of long-term patients in general hospitals. Complaints about the inability to get patients out of acute beds were as common in the 1980s as they had been the decade before.[49] Nonetheless, massive reliance on private nursing homes was one of the most characteristic features of the American approach to chronic illness. As the leading historians of nursing homes expressed it: "In this manner, a particularly American solution to long-term care needs emerged. It rested on the traditional distrust of a large and activist government, interest group politics, and the continued need for care otherwise unavailable. By the 1970s, institutionally based structures would become the standard around which alternatives would be created. These alternatives were assessed principally in terms of their cost-saving potential."[50] The chief of these alternatives was homecare.

Homecare

Homecare was a charitable enterprise in the nineteenth-century United States; by 1909, nearly six hundred organizations were sponsoring the work of visiting nurses. As we saw in chapter 1, a growing number of life insurance companies were in the early twentieth century offering it as a benefit. But insurance activity in this field began to wane in the 1940s as elderly, chronically ill people became the major users, and it turned out that homecare made little difference in terms of recovery or longevity. In 1953 both Metropolitan Life and John Hancock ended their visiting nurse programs. Nonetheless, health reformers after World War II gradually came to see homecare as a less expensive and more effective alternative to hospitalization. As K. Buhler-Wilkenson and others have pointed out, any savings were usually the result of shifting the burden of care to unpaid family members, more often than not women.[51] A more recent study has emphasized that low-wage labor of poor women has also been instrumental in keeping down homecare costs so that it is not merely a privilege of the very rich.[52]

The gold standard for homecare was the hospital-based homecare program. In Syracuse, New York, in the early 1940s, a demonstration was set up to deal with the recurrence of illnesses (predominantly chronic) that followed discharge from hospitals. It was claimed that this hospital-operated homecare system had produced savings due to avoidance of unnecessary hospitalization equivalent to more than three times the cost of the experiment.[53] The emblematic model for such care was a program organized at the by now iconic Montefiore Hospital in New York City in the late 1940s by E. M Bluestone, who became a much cited guru in this field;[54] this model was based on centralized hospital management of a variety of services, including nursing care, social work, and housekeeping. Such plans were referred to as "coordinated homecare programs." The aim was to establish a continuum between hospital care and the home through "the continuation of comprehensive health services in the home."[55] The level of medical care was expected to be of hospital quality. Although technically distinct from medical services, homemaker services were ideally considered part of the program. In 1948 the New York City hospital system established a hospital extension system on the Montefiore model. By 1950, sixteen of the city's thirty-five hospitals had homecare programs.[56]

Homecare was a centerpiece of the strategy recommended by the CCI, which, in collaboration with the PHS, surveyed eleven plans in a variety of locations.[57] While these were deemed to be effective, the study regretted that such programs were relegated to indigent people. "Patients of all economic groups, especially those with long-term illness, need coordinated services at home during some phase of illness."[58] Everyone seemed convinced that healthcare costs would diminish if homecare became more widespread. This assertion, however, was never demonstrated with convincing data and such programs developed slowly. The PHS, in conjunction with the AMA and Blue Cross and Blue Shield, conducted a series of surveys and found that the number of these programs more than doubled between 1960 and 1963, from only 33 programs in 18 states, to 70 programs in 23 states. While this sounded impressive, the numbers involved were small; in 1963 programs cared for only 550 patients.[59]

But federal funding through Medicare, Medicaid, and Title III of the Older Americans Act completely transformed the landscape by providing new funds for homecare.[60] Medicare gave priority to acute illness, and homecare, like nursing homes, seemed a way of getting people out of hospitals more rapidly. Much of the regulatory activity of the period sought to increase the supply of homecare and extend its use by easing restrictiveness of benefits. Within a short time, Medicare became the single largest payer of homecare services and was defining the conditions of its use. By the late 1960s, from 6 to 7% of general hospitals had homecare programs, more than double the number in 1965.[61] Some hospitals set up transition programs to get patients used to homecare.[62] By the 1970s, homecare was perceived as a viable alternative to nursing homes for older people. The trend seems to have been for multiunit programs serving groups of hospitals. In Chicago, a homecare program administered by the local visiting nurse association was set up with six participating hospitals providing continuity of care between hospital and home; services included bedside care as well as physical, occupational, and speech therapy for patients.[63] For patients unable to live at home, Baltimore City Hospitals set up a system of single-patient community care homes; an individual or family would take a patient into their home and in exchange for remuneration take full-time care of him or her.[64] The Older Americans Act provided a variety of home and community services for frail elderly individuals. Although complaints of restrictiveness continued to be heard, after 1975 the number of programs rose steadily. Homecare was of course not exclusively about chronic illness or older people, but elderly, chronically sick individuals accounted for a large proportion of those served.

There were, however, several axes of tension and conflict running through homecare. Most basic was the tension between the Medicare model designed to provide short-term, post-hospital care, mainly through skilled medical services (though incidental supportive care was permitted) and the Medicaid model with more expansive services, including skilled and non-skilled, supportive and preventive, and curative care to low-income chronically ill people who required no prior hospitalization. (Both, however, resisted paying for domestic and homemaker services.) Homecare was also vulnerable to political considerations. In the 1970s studies found long-term homecare services in California and Massachusetts to be in decline due to austerity measures applied by Medicare and Medicaid between 1967 and 1970.[65] Still, homecare costs roughly doubled between 1975 and 1978 as both Medicare and Medicaid sought to stem the even greater costs of institutional care. Overall, the thrust of legislative and administrative changes during this period was to increase use of homecare by easing restrictive regulations, expanding benefits, removing limits on number of visits, ending requirements for prior hospitalization, eliminating or reducing deductibles, and permitting reimbursement to proprietary homecare agencies.

Nevertheless, the rapidly increasing costs and the absence of evidence demonstrating the benefits of homecare gradually generated a more critical outlook.

Studies found that broadened coverage of homecare services did not improve ability of clients to function, lower rates of institutionalization, or reduce costs of care.[66] What made growing costs even more difficult to accept was the spread of profit-making agencies, frequently charged with corruption and mismanagement. "By the end of the 1980s, one-third of all Medicare-certified home care agencies were for profit, and by 2003, nearly half were proprietary."[67] Despite serious doubts about the direction it had taken, homecare became an accepted part of the healthcare landscape. By 1987, 5.9 million Americans of all ages were receiving homecare services, at a cost of $22.3 billion annually, with the federal government paying about 46%. The remainder was privately financed (out-of-pocket and private insurance).[68] With so much money at stake, hospitals moved into the home healthcare business on a large scale. In 1980, 510 (8.1%) of AHA-recognized hospitals had a homecare department. A decade later, 1,900 hospitals (31% of the total) provided home health services.[69] While this gave greater legitimacy to homecare within medicine, hospital administrations tended to promote the post-acute, skilled-care model (Medicare) at the expense of the maintenance and support services model (Medicaid).

There has developed a general consensus that homecare has become a significant reality in the United States but that it is inadequate in relation to need and likely to remain so. In the view of its most authoritative historian, it cannot be the basis of chronic care for many reasons: confusion about goals (rehabilitation versus maintenance), tension between medical and social perspectives, doubts about the reality of the cost-savings that often justify it; and the disjunction between legitimate needs and fiscal realities. According to K. Buhler-Wilkerson, "the needs of the chronically ill have constantly confounded our ability to develop a comprehensive or informed system of community-based care . . . The private, unseen, and seemingly uncontrollable nature of caring for the sick at home, combined with the open-ended nature of chronic illness, make the institutionalization of home care in the continuum of health care essentially untenable in the context of political, social, and economic realities, cultures, and incentives in the United States."[70]

Rehabilitation

Rehabilitation was at the center of reformers' optimism about controlling chronic disease. In the mid-1950s Hill-Burton funds became available to build or renovate rehabilitation centers whether in hospitals or as stand-alone institutions. Nursing homes increasingly had rehabilitation facilities as well. Hospital publications described numerous initiatives. An addition to the North Hospital in Seattle had several rehabilitation units, including a stroke care program and a residential language rehabilitation program. A Detroit rehabilitation center planned a $3.5 million addition to double its bed capacity.[71] St. Joseph Hospital in Chicago developed a program of workshops to teach rehabilitation techniques to nurses employed in nursing homes.[72] Franklin County Public Hospital in Greenfield, Massachusetts, organized in 1960 an activities of daily living (ADL) program within a special hospital unit geared to increase the capacity of elderly patients to care for themselves.[73] A chronic disease hospital in Baltimore did a major study to see if early rehabilitation provided advantages for stroke victims. (Results were mixed. More patients died with early rehabilitation but survivors achieved a higher rate of improvement.)[74]

As the above examples illustrate, rehabilitation could mean many things. Early in the century it was geared to getting people back to work. The Industrial Rehabilitation Act passed in 1920 created a modest program with a vocational orientation. While including a medical component, the primary goal was "training around" the disability rather than seeking to lessen or eliminate it. By the 1920s and 1930s, a field that came to be called physical medicine emerged. At first focused on diagnosis and treatment of disease, with emphasis on orthopedic and rheumatologic problems, it became a major priority during World War II due to the work of military physicians like Howard Rusk and Henry Kessler. There was thus from the beginning a tension between two distinct programming strands, medical and vocational, which were frequently under separate administration but which remained linked by overlapping client populations. The gap between these two strands widened as medical rehabilitation expanded its scope to the sequelae of many illnesses that resulted in lost function. In the process, rehabilitation medicine moved from military to civilian life and well beyond physical medicine, defining itself as a comprehensive, multidisciplinary enterprise, including therapists of every sort—nurses, psychologists, social workers, as well as physicians—concerned with the functional, psychological, and vocational capacities of the "whole man."[75] A committee chaired by philanthropist and former presidential advisor Bernard Baruch (whose father had been medical director of Montefiore Hospital) prepared a report that formalized the ideal of comprehensive rehabilitation. The program initiated by Howard Rusk at New York's Bellevue Hospital embodied this model and achieved iconic status during the postwar decades. A 1961 survey of rehabilitation facilities funded by Hill-Burton monies found that of the 1,724 institutions reported, 436 were defined as offering comprehensive rehabilitation programs, with "formal services in the medical, psychological, social and vocational areas."[76]

The creation of Medicare in 1965 was a major turning point for rehabilitation medicine. Under its provisions, medical rehabilitation services for patients over the age of 65 were covered as part of hospital inpatient services. This and other measures of the period allowed for the incorporation of rehabilitation as an integral aspect of post–acute care for major chronic illnesses and further contributed to the expansion of rehabilitation programs and services. Many hospitals signed agreements with rehabilitation centers, while others established their own. Both Hill-Burton and the 1966 Amendments to the Vocational Rehabilitation Act provided increased funding for rehabilitation facilities. Expansion continued into the 1970s; the 1972 Amendments to the Social Security Act extended Medicare benefits to people with disabilities under the age of 65 who were already receiving Social Security Disability Insurance and this coverage was extended still further by the Rehabilitation Act of 1973 and subsequent amendments.[77]

The goals of rehabilitation also expanded. Many within the field, along with members of the growing disability rights movement, advocated for "independent living" services to be included as part of rehabilitation, with a view to returning individuals to functional autonomy in the home and community. Several pieces of legislation that aimed at independent living rather than vocational training were defeated or left out of bills that were passed. The 1978 Amendments to the Rehabilitation Act of 1973 extended coverage to people with disabilities under 65 who did not have vocational goals, provided funding for comprehensive rehabilitation programs and facility construction, and created two Independent Living Programs.[78]

The goals of the field extended in other directions as well. From the 1940s through the 1960s, while amputations and orthopedic problems continued to be major focuses of rehabilitation, the fastest-growing programs targeted mentally ill and intellectually challenged people. By 1970, approximately 25% of rehabilitation research was on mental illness, fueled in part by the deinstitutionalization movement and the redefinition of drug and alcohol addiction as mental illnesses.[79] Throughout the 1960s, discussions of the future of rehabilitation invariably touched on the need to deal with the aged disabled population. Medicare greatly increased possible funding and studies were launched to explore the potential of rehabilitation programs for this age group.[80] Nonetheless, it was only gradually, and partly in tandem with the rise of geriatrics in the United States, that older people came to be considered as fit candidates for rehabilitation. Chronically ill aged individuals, with multiple comorbidities and an uncertain life expectancy, did not make for ideal rehabilitation candidates, despite studies suggesting that rehabilitation could sometimes be effective;[81] it has been argued that cultural attitudes about old age as a period of "natural" decline and loss of function have played a role in the reluctance to aggressively pursue rehabilitation for elderly people.[82]

Throughout the 1970s and 1980s the number of institutions offering rehabilitation services in one form or another grew steadily. According to a survey conducted by the American Hospital Association, from 1975 to 1980 the number of inpatient facilities remained fairly stable at about 415 units mainly concentrated in urban centers, whereas those providing outpatient services more than doubled, from 610 in 1975 to 1,235 in 1980. By 1990, the number of outpatient services had increased to 2,830. Viewed from another perspective, 6.5% of recognized hospitals had comprehensive rehabilitation inpatient services in 1980 while nearly 20% had outpatient services; a decade later, the figure was 46% for outpatient services. (Inpatient services disappeared as a category in 1990, seemingly broken down into more specialized units.)[83] These figures include only AHA-registered hospitals, and exclude other institutions like the vocational rehabilitation facilities that offered medical services and skilled nursing homes with rehabilitation units. There were, in addition less comprehensive options; in 1980 one-third of all registered hospital institutions had occupational therapy departments and more than three-quarters had physical therapy departments. Of the inpatient rehabilitation units, three-quarters were located within acute-care or general hospitals concentrated mainly in major urban areas.[84] In this one area, at least, the ideal of a comprehensive approach to chronic illness centered in hospitals was achieved.[85]

One reason for the rise in the number of hospital rehabilitation institutions from 1980 to 1990 was the Omnibus Budget Reconciliation Act of 1980 that expanded Medicare reimbursements and resulted in a proliferation of private rehabilitation centers and corporations.[86] After much lobbying, rehabilitation facilities were in 1983 exempted from a prospective payment system (PPS) in which Medicare reimbursed hospitals on the basis of what specific diagnostic-related groups "should" cost.[87] One result of this exemption is that criteria became more stringent. In order to obtain PPS exemption as an inpatient rehabilitation facility, at least 75% of the patients in treatment at the facility had to fall into one of the following diagnostic categories: stroke, spinal cord injury, congenital deformity, amputation, major multiple trauma, fracture of the femur, brain injury, neurological disorders, burns, and polyarthritis; and patients were required to receive a minimum of three hours of intensive physical or occupational therapy, five days per week.[88]

Acute-care facilities without the PPS exemption had a strong incentive to discharge patients to rehabilitation centers as early as possible. Consequently, between 1986 (when PPS was fully integrated) and 1994, referrals and admissions to inpatient rehabilitation facilities more than doubled, and the number of Medicare-certified rehabilitation hospitals/units increased by 87%, from 545 to 1,019. From 1990 to 1993 alone, Medicare reimbursement to rehabilitation facilities grew from $1.9 billion to $3.7 billion.[89] There was, however, a negative consequence to this growth. Under the new payment system there was a reported increase in inappropriate referrals and in patients who required continued acute care or suffered additional complications from unresolved health conditions. This forced rehabilitation facilities to offer increased medical services and sometimes take on the management of continued acute conditions.[90]

Rehabilitation institutions have probably been the most successful and least controversial of the chronic-care institutions so far discussed. But while they have received political and administrative support, their place in mainstream medicine has not been so clear. Although physical medicine became a specialty in 1947 and "rehabilitation" was added to the specialty name a few years later, it has not enjoyed high status within the profession and until the 1980s had great difficulty attracting medical graduates.[91] Although this situation has improved due to increased funding and growing numbers of facilities, there remains considerable speculation about the reasons for the low status of the specialty: a holistic medical perspective that is not shared by most doctors; failure to pursue solid research and provide real evidence of efficacy, with the paucity of clinical trials the latest manifestation of this charge; the amorphousness of a field that focuses on no particular organ or pathology and utilizes radically different and specialized techniques for different conditions and thus has persistent difficulties of self-definition; and competition from orthopedists, rheumatologists, physical and occupational therapists.[92] For all that, physical medicine and rehabilitation, while not a large specialty, currently has more than nine thousand practitioners. They do not earn the highest specialist incomes but neither do they earn the lowest.[93] As we shall see in the following chapter, rehabilitation is an integral component of many state and local chronic disease programs.

Conclusion

Various provisions for long-term care were developed in the decades following the report of the CCI. The ideal of hospital-based care was not totally abandoned but turned out to be far too expensive to be developed fully, except in the case of rehabilitation and for a time skilled nursing units. Hospitals have, in fact, outsourced many functions in addition to chronic care for similar financial reasons. The United States thus has a huge nursing home industry and a significant home-care sector. Both are increasingly run for profit, and both are saddled with serious problems. Rehabilitation institutions have faced less criticism but have had to moderate their aspirations as they deal increasingly with aging patients. While the United States was unique in the way that it framed the problems of long-term care as chronic disease, the difficulties that it faced were not essentially different from those confronting the European countries to be discussed in the final chapters of this book. What differed was the policy framework.

CHAPTER SEVEN

Public Health and Prevention

When Lester Breslow first applied for a job with the California Health Department in 1946 in order to work on chronic disease, he was initially turned down because the head of the department had no interest in this issue. The situation, however, turned quickly around. Breslow was eventually hired for another job and soon emerged as head of the state's newly created Chronic Disease Service that immediately took responsibility for a federally funded cancer control program "in recognition of the fact that cancer is but one of many chronic diseases which may be subject to the same general public health approach." Soon after, he also became chairman of the American Public Health Association (APHA) Committee on Chronic Disease and Rehabilitation.[1] In the years that followed, governmental bodies, federal, state, and local, organized around the chronic disease problem. The process intensified gradually until the Commission on Chronic Illness (CCI) made it a major priority by the mid-1950s. Indeed, the turn to chronic disease was often depicted in journals and meetings as a major turning point and engine of transformation for public health as a field.[2]

As in other domains, the new emphasis on chronic disease generated a number of strains within public health institutions. The first had to do with the CCI's policy of integrating all aspects of chronic disease activities into comprehensive programs. There was certainly considerable support for this view. An APHA publication of 1960 asserted: "The most effective basis for organizing chronic disease activities is to integrate all phases and stages of attack—prevention, early detection, treatment, rehabilitation—into programs that recognize the common denominators among chronic disabilities in various disease categories, age groups, and sources of service or funds."[3] This, however, was easier said than done. The most significant barriers to this ideal were the appeal of categorical *disease-based* programs—especially but not exclusively cancer—and the numerous fragmented programs and agencies that had accumulated over the years in every state and local jurisdiction. A second potential obstacle was the emphasis on biomedical research, which attracted more public and private funds than practical public health programs. But the most fundamental problem, although seldom acknowledged, was the huge variety of tasks making up chronic disease policy; caring for bed-ridden invalids required very different resources, personnel, and approaches than preventing or finding disease in healthy populations. The Public Health Service (PHS) and other national agencies did their best to bring leadership and direction to the smorgasbord of existing programs, but even at this national level, pressure to develop distinct categorical programs remained intense.

Activity at the National Level

The PHS was at the forefront of the federal campaign against chronic illness. The Bureau of State Services of PHS had a Division of Chronic Disease, or some similarly oriented agency, from 1949 on. In 1967 it was absorbed by the Bureau of Health Services, and became the National Center for Chronic Disease Control. Consistent with the recommendations of the CCI, the division/center emphasized comprehensive approaches to chronic illness, while recognizing that disease-specific programming was effectively unavoidable. It thus administered its own categorical programs for cancer, diabetes, arthritis, and heart, neurological, kidney, and sensory diseases, and by the mid-1960s, respiratory problems, smoking, and health. The division/center organized training and education programs for nurses and other public health professionals, surveys of existing services and health needs, and demonstrations for early detection, diagnosis, treatment, and rehabilitation of various diseases. It conducted research, and operated its own units devoted to rehabilitation and environmental health, and developed a program for the establishment of kidney dialysis centers.[4]

As gerontology and services for the aged became increasingly identified with chronic illness, the division in 1963 formed a Gerontology Branch, concerned with the "aging and aged." It turned out, however, that the scope of branch activities was substantial since "aged" referred to those 65 and older, but "aging" was defined as starting at 45. The core responsibility of the Gerontology Branch was the promotion of a program of health maintenance for people over 45, which included periodic examinations, multiphasic screening, public education projects about health maintenance (including exhibits, literature, and films), publicizing available services (counseling, referral, and follow-up), and coordinating community health services. Yet another division, the Nursing Homes and Related

Facilities Branch, was created slightly later to oversee projects relating to long-term care.[5]

Such federal agencies compiled information about the chronic disease problem and sought to overcome the many weaknesses associated with a primary research tool, the population survey. The PHS had been directly involved in the morbidity surveys conducted by the CCI, and its chronic disease unit initiated a national longitudinal study of childhood obesity and adult mortality. The National Health Examination Survey (NHES) conducted by the National Center for Health Statistics unfolded in three phases from 1959 to 1970, with the explicit goal of collecting information on chronic illness prevalence. Phase I of the NHES, conducted from 1960 to 1962, focused on chronic disease, especially cardiovascular disease, arthritis, and diabetes, in individuals ages 18 to 79, whereas Phases II and III looked at children and adolescents. While the survey was effectively a descriptive endeavor, it also served as a mass-screening project since a major goal was to detect previously undiagnosed chronic disease.[6] In the early 1960s the National Health Survey also carried out projects in collaboration with the New York Health Insurance Program and the California Kaiser Permanente Medical Group to test and improve the methodological dimensions of the surveys and explore the discrepancies between interview and health examination data on chronic disease. The PHS and the University of Michigan in 1963 organized a symposium on genetics and chronic disease, with the goal of confronting public health survey methodology with that of population genetics. Other symposia it sponsored focused on specific diseases like emphysema.[7] In the 1970s attention turned to the role of nutrition in health and disease starting in 1971 with The National Health and Nutrition Examination Surveys.[8]

Along with nationwide information collection, research, and planning for chronic illness, the PHS provided funds as well as professional and technical assistance to stimulate programs and activities at the state and local levels. The goal was to put into practice the best existing knowledge for the control of specific diseases and to evaluate the technology. Activities supported included case-finding programs, model projects and community programs in rehabilitation, public health films, and testing regimens in model outpatient clinics. Such programs were in the end about keeping the medical and other healthcare professions abreast of the newest techniques in all these domains. While such programs often targeted specific diseases, the PHS prided itself on not limiting itself to any disease category or age group.[9]

The PHS in 1950 administered $43.1 million in projects, almost all in the form of general grants, with small amounts directed to specialized cancer control, mental health, and heart disease programs. In 1966 the PHS handed out $138 million, mostly in the form of special projects grants. It is not easy to determine exactly what programs centered on chronic disease but using the most conservative definitions and excluding programs for tuberculosis and mental illness, chronic diseases accounted for about 40% of the total monies spent. (Adding mental illness raises the figure to a little less than 50%.) But it is worth noting that grants were primarily categorical. Funding for cancer was $17.4 million; for heart disease, $9.5 million; and for mental health, $26 million.[10] There were smaller appropriations for conditions like neurological and sensory diseases. A little more than $12 million was devoted to the general category "Chronic illness and the aged" with another $9 million directed to home health services. Although the emergence of this comprehensive funding category was certainly significant, general chronic disease funding lagged far behind that for disease-specific programs. The Partnership for Health Act in 1966 eventually replaced nine categorical grants, including those mentioned above, along with home health services, tuberculosis control, and dental services, with a block formula grant, on the assumption that this would increase state and local control and lead to more efficient use of federal funds. It is unlikely, however, that block grants encouraged more general chronic disease programs since ongoing state and local categorical programs had to be sustained and federal funds for public health did not increase significantly during the Nixon years.[11]

Federally Supported Research

These figures do not include what was by far the largest federal investment in the chronic disease problem: biomedical research and particularly that of the National Institutes of Health (NIH). We saw in chapter 3 that the PHS hygiene laboratory became the National Institute of Health in 1930. The National Cancer Institute (NCI) was created in 1937. During the war and postwar years, a powerful "research lobby" (whose leadership is usually attributed to Mary Lasker) was able to convince Congress and several presidents of the centrality of research. One result was the creation of the National Science Foundation in 1950; another was the revitalization and restructuring of the NIH.[12] In the years following the war, the NCI was absorbed into the NIH (pluralized as "Institutes" in 1948). In response to pressures from voluntary health organizations, Congress soon added institutes for mental health, dental diseases, and heart diseases. By 1960, there were 10 institutes and, by 1970, there were 15. The total NIH budget rose from

$8 million in 1947 to $70 million in 1952 and to $1 billion by 1967.[13] Research was the one means of satisfying growing popular demand for good health in a way that all key medical and political groups could support.

The leaders of the NIH tried but failed to withstand the pressure to create disease-based institutes, not because they believed in a comprehensive chronic disease program but because they were committed to fundamental biological research.[14] The NIH leadership tried to minimize the narrowness of disease-specific structures by emphasizing that fields like metabolic or neurological disorders, heart disease, and cancer were in fact very broad. "Our philosophy is that an understanding of the basic disease processes will be impeded unless our investigators have elbow room to explore the fundamental relationships among body systems and the fundamental biochemistry and biophysics of protoplasm."[15] This was not mere rhetoric. In spite of the NIH's disease-based structure, its director, W. Sebrell, did research on the role of nutrition in chronic diseases generally and the NIH developed a reputation for emphasizing fundamental research.[16] This did not always please congressional backers and research lobbyists, who took a less expansive view and frequently sought practical remedies to specific diseases.

The diseases studied in the NIH were almost exclusively those categorized as "chronic" and this fact was trumpeted in institutional publicity. A press report on the opening of the NIH's Clinical Center made clear that the "primary function of the new research center is to study those diseases that are responsible for the greatest number of deaths and disabilities such as cancer, heart diseases, mental illness, and metabolic disorders."[17] When Dr. C. J. Van Slyke, associate director of the NIH, was questioned on television in 1953 about the reason for establishing the NIH research program, he responded by citing the urgency of finding solutions to chronic disabling diseases.[18]

The federal government was not alone in financing chronic disease research. By the 1950s, pharmaceutical companies in the United States and abroad were becoming aware of the economic potential of research that might yield therapeutic substances that did not cure but that would be purchased for many years.[19] Chronic-care hospitals like Montefiore, Goldwater, and Jewish Chronic Disease Hospital of Brooklyn had for some time had research institutes. Universities were also responding to the new priorities and new sources of research funds. By the 1960s, most universities were doing research on one or another chronic illness, with the majority taking place in discipline-, disease-, or organ-based departments or institutes. But there were more comprehensive research institutions as well. The University of Buffalo had a Chronic Disease Research Institute that published actively throughout the 1950s and 1960s. The Johns Hopkins School of Public Health had a department of chronic disease headed by Abraham Lilienfeld, and the University's Medical School had the Joseph Earl Moore Clinic specializing in clinical research on a variety of chronic diseases. In 1967 UCLA established an Institute of Chronic Disease and Rehabilitation.[20] One could also mention the Texas Institute for Rehabilitation and Research at Baylor College of Medicine founded in 1959.

Still, maintaining a comprehensive orientation was difficult. The *Journal of Chronic Disease*, which tried to maintain both a multidisease and a multidisciplinary orientation was from the beginning viewed with suspicion. The eulogist of one of its founders described the journal in these words: "At the moment this new publication is successful and seems to be bridging gaps between orthodox scientific and clinical medicine on the one hand, and gerontology, preventive medicine and public health on the other." The eulogist went on to note that the founder's friends "all" advised him against taking a position of editor. [21] The journal nonetheless had a respectable run before finally abandoning its multidisciplinary orientation in 1988 to become the *Journal of Clinical Epidemiology*, which made considerable sense since epidemiology had largely turned to the study of chronic diseases.

Persistent political pressures had two effects on research efforts. First, they intensified demand for disease-based research, especially on cancer. Second, there was consistent pressure for practical clinical solutions, for cures rather than basic research. Occasionally, it led to calls for research institutions to take a more practical public health role. In 1971, again pushed by the cancer lobby led by Mary Lasker, President Richard Nixon signed the National Cancer Act that declared a "war" on cancer and that significantly expanded the role of the National Cancer Institute and gave it greater autonomy from the NIH. Its research role was appreciably expanded by new funding (more than $5 billion by end of 1978), making it "the largest single biological research offensive that the U.S. has known." This led to new funding throughout the NIH, much of which was spent on "mission"-oriented research contracts.[22] Additionally, the director of the NCI was given authority to plan and coordinate a national cancer program that included initially fifteen cancer centers and state and local cancer-control programs.[23] In the first years of the program, most of the emphasis was on research, with the major controversies revolving around the relative weight of basic versus more practically oriented inquiry and of research on cancer versus that on other diseases. There was also pressure to actually implement the practical programs that were part of the 1971 law. But in a five-year review of the National Cancer Program, the only reference that its director made to such practically oriented activities was to demonstration programs in "those areas where beneficial technologies exist but where their acceptance and utilization is not widespread. This is an area that troubles a great many scientists because they are concerned that the entry of the NCI into the service area, ultimately results in the diminishing of its resources for research."[24] The push for practical action affected more than the NCI. The NIH was for several years in charge of the Regional Medical Program (to be discussed below). In 1972 Congress established the National High Blood Pressure Education Program and located it as well in the NIH.[25]

Activity at State and Local Levels

In 1944 only two states had chronic disease units. By 1956, thirty-one state health departments had such units. The first annual meeting of directors of state chronic disease programs took place in September 1955. Twenty-seven states and Hawaii sent representatives.[26] In 1960 the recently formed Association of State and Territorial Chronic Disease Program Directors held a symposium and lecture series for public health physicians, nurses, educators, and consultants. That same year, the American Public Health Association initiated a continuing education program for public health workers in thirteen western states; it included, among other subjects, special seminars and courses devoted to chronic disease.[27]

Nonetheless, it was common, perhaps the rule, for state or local health departments to offer a range of disconnected services relating to different diseases. The spread of dedicated state chronic disease departments does not appear to have extended to local health departments, where most direct action took place. In a 1956 survey of local health departments, 60% reported that chronic diseases were considered a major responsibility, but only 10% had administrative entities specifically devoted to them. Furthermore, "chronic disease activities" could refer to anything from a cancer registry to a nutrition program to home nursing care; twelve states had programs for Pap tests, nineteen for blood sugar, and nine for blood cholesterol testing.[28] The existence of a state division for chronic disease in no way guaranteed the comprehensive local programs that national agencies were advocating. One of the reasons was that federal funds were categorical until 1968 and by then block grants had to be used to maintain existing programs. Mental illness was generally administered separately, partly because programs and institutions for it already existed and partly because even at the national level it was treated as a distinct social and administrative problem. The division of services between departments of health and departments of welfare further promoted fragmentation. By 1976, little had changed even though twenty-nine states had statutes specifically authorizing chronic disease control as the responsibility of state or local health departments.[29]

Some state health departments directly organized and managed projects in local health departments. In others cases, county and local health departments operated relatively independently, with the state overseeing funding, licensure, loan of personnel, or standards. Of the states in 1976 with statutes authorizing chronic disease control, fourteen made chronic disease the exclusive responsibility of the state health department, twelve made it the joint responsibility of state and local agencies, and three mandated chronic disease control as a specifically local responsibility.[30] State health departments that were directly responsible for at least some local health agency activities included Connecticut, Maryland, Minnesota, Tennessee, and Virginia. States where local city or county health departments operated relatively independently, with the state playing a consulting or advisory role, included California, Illinois, Maine, Massachusetts, New York, and Ohio. Joint operation of programs in one form or another was observed in Alabama, Missouri, New Jersey, and Oklahoma. The states best known for their chronic disease activities were California, Maryland, and New York.[31]

The California Department of Public Health set up a cancer registry in 1946 and formed the California Chronic Disease Service (later the Bureau of Chronic Diseases) that same year; its early work was for the most part focused on cancer and funded with federal monies. Besides the registry, programs included early detection, epidemiological studies, surveys of existing services, and the training of professional personnel. The prevailing public health philosophy in the state emphasized the relative autonomy of local departments, and so activities varied from one county to another. In 1950, however, the California Conference of Local Health Officers adopted the state's Chronic Disease Control Program Guidelines produced in order to standardize local efforts. An early focus of the state public health department was the collection of morbidity data and the systematic evaluation of existing services and programs available for chronic illness. In 1954 the department of public health, with support from the NIH, conducted the California Morbidity Survey, which focused heavily on chronic disease. The same year the department carried out a survey of local health department activities.[32] Lester Breslow became a tireless spokesman and major figure in the national chronic disease movement. He was an enthusiastic supporter of multiphasic screening, which gained wide popularity within the state.[33] Despite strong left-wing sympathies that suggested the need for social change, he believed deeply in programs of education to modify individual behavior. But he had as well an ongoing interest in environmental and occupational health issues. He was a key figure in drawing up in 1960 air quality standards for the state.[34]

The New York State Department of Health began cancer reporting in 1911. During the 1940s, it organized a commission that, among other things, prepared a comprehensive plan for dealing with chronic illness.[35] By 1956, the department had a Bureau of Chronic Disease and Geriatrics. It immediately began to organize diabetes and glaucoma screening and expanded to multiphasic screening in 1959. In 1962 the department introduced a 5-year projected plan that included the development of 10 multiple disease screening programs, 6 to 12 health maintenance clinics for aged individuals, 12 new homemaker programs along with a similar number of coordinated homecare programs, and 20 information and referral centers for chronic disease.[36] Nonetheless, the majority of New York's programs for chronic disease were focused on specific diseases, notably cancer, coronary and cardiovascular disease, and chronic respiratory disease. One exception was the statewide medical rehabilitation program for crippled children, a state-run but locally administered program that in 1964 expanded to cover cystic fibrosis, diabetes, chronic renal disease, chronic asthma, and leukemia.[37] As in California, concerns over air pollution increasingly translated into action in the 1960s, in the form of research, stricter emission standards, and monitoring activity by public health agencies. Chronic respiratory diseases were one focus of this activity, but air pollution was considered a factor in a variety of chronic conditions. Other states followed the lead of California and New York on this issue.[38]

Maryland also took early action against chronic disease. In 1945 the Maryland Medical Care Program made the health of indigent persons a responsibility of the department of health rather than the department of welfare. Over the course of the 1940s and 1950s, several new state-run chronic disease hospitals were constructed.[39] State and local authorities cooperated with the PHS in conducting a wide range of morbidity studies in Hagerstown, thus continuing and extending the original work of Edgar Sydenstricker. Just before World War II, a pioneering longitudinal chronic disease study was begun in East Baltimore as a collaboration between the state department of health, the PHS, the Milbank Memorial Fund, the Johns Hopkins School of Hygiene and Public Health, and the Baltimore City Health Department. The expertise thus generated led to the much larger CCI study of Baltimore that became volume 4 of the commission's report. Baltimore continued to be a hub for chronic disease research, concentrated mainly at Johns Hopkins, producing various follow-up studies to the CCI's report over the years.[40] The department of health's Chronic Illness Program introduced, among other activities, a referral system for various health and community services and a mobile health unit that provided screening for chronic diseases, specifically glaucoma and diabetes.[41]

In New Jersey, the 1952 Prevention of Chronic Illness Act passed by the state legislature called for the formation of a Division of Chronic Disease in the department of health and an Advisory Council on the Chronic Sick, while setting aside considerable state funds for grant-in-aid programs to local agencies. The state's activities are less well known than those of New York or California but are singular for the close involvement of state public health officials with local hospitals and voluntary agencies. Grant-in-aid funds were designed as start-up money for local projects; in the initial phase the state provided funds and worked closely with the recipient agency, with the project expected to achieve functional independence within three years. In the initial planning, $185,694 was earmarked for a variety of projects, with the bulk going toward early detection (56%) and rehabilitation (25%).[42] These state funds were in addition to pre-existing federal grant-in-aid money for heart disease, cancer, and tuberculosis programs. According to the commissioner of the state department of health, "There has been continuously a deliberate special emphasis on strengthening generic or more generalized services in the prevention of chronic illness program, at the same time that the traditional disease centered programs have been expanded." By 1964, $1,692,000 had been distributed by the state's Division of Chronic Illness, independent of categorical federal grants.[43]

A similar program was operated by the Massachusetts Department of Public Health, which supported chronic disease services offered by voluntary agencies through a small-grants project program starting in 1961. By 1966, thirty-five projects had been initiated.[44] Local authorities also provide examples of impressive achievement. In Memphis, Tennessee, the county health department in cooperation with the University of Tennessee and City of Memphis Hospital started in 1963 a decentralized network of chronic disease care centers that continued to operate into the 1980s. These clinical centers, staffed by public health nurses and supported by local physicians, provided patient education, medical care, and referral services for a variety of chronic conditions, primarily diabetes, hypertension, and cardiac disease.[45] In the late 1940s Richmond, Virginia, initiated a largely independent local chronic disease program that organized multiphasic testing pilot programs, homecare and rehabilitation services, and continuing education and training in chronic disease for physicians and medical students.[46]

There was much activity in certain states and cities from the 1950s to the 1970s. Still, the overwhelming impression is that most local and state agencies managed collections of unconnected, mostly categorical programs with an emphasis on cancer, heart disease, and diabetes. This is true even of New York and California, considered models of coordinated planning. It could be argued that this was an exigency of practical program implementation, that categorical programs made sense despite the desire to unify programming. Indeed, tension between the generalized, coordinated "chronic disease" approach and the disease-specific models that predominated in reality was a matter of concern for administrators. In their 1957 survey of local chronic disease activities, J. N. Muller and E. B. Kovar described "a great many services, from the health department and from other agencies, but no chronic disease program" and posed the fundamental question: "Is there really a common thread which binds the chronic disease services together, but which does not tie them to services relating to 'non-chronic' disease? . . . What makes a program a program, rather than a collection of services? What shall we say is the purpose of a local chronic disease program? How does its purpose differ from a cancer control program?"[47]

By the end of the 1960s, a great many bureaus and divisions of chronic disease had sprung into existence but comprehensive chronic disease programs had largely failed to materialize, or had materialized briefly only to disappear. This situation did not change dramatically during the next decades. In 1986 the Centers for Disease Control (CDC) and the Association of State and Territorial Health Officers jointly organized the First National Conference on Chronic Disease Control and Prevention. Much was discussed and agreed to, but no consensus was achieved regarding the relative value of categorical programs versus comprehensive programs or even about what the terms *categorical* and *comprehensive* actually meant.[48]

There were a number of ways that the tension could be addressed if not resolved. The simplest was simply putting many categorical disease programs under the same roof and administration. Local health agencies clearly did that as did chronic-care hospitals holding different kinds of patients and services. But this fell short of the ambitions of reformers. A deeper form of coordination or unity was provided by common or at least related techniques. Rehabilitation centers locating different illnesses and procedures in the same building or set of buildings had a degree of technical and conceptual coherence that balanced the variety of their practices. Nutrition was another arena that cut across specific disease categories; diabetes, obesity, and heart disease were most obviously concerned, but benefits from proper nutrition applied to most chronic conditions. Many rehabilitation centers had little to do with public health agencies, however, while nutrition programs were most often not managed by chronic disease divisions of public health departments.[49]

Multiphasic Screening

The public health measure that best embodied the collective approach to multiple chronic illnesses during the 1950s and 1960s was multiphasic screening. Screening itself was not new. It had been applied early in the twentieth century to soldiers and schoolchildren. Insurance companies, businesses, and public health agencies had promoted periodic physical examinations early in the century. Although not popular with the public or physicians, examinations never disappeared and gained new life in the 1950s through prepaid group practice plans, most influentially, the Kaiser Permanente Health Plan in California, which began to offer periodic examinations in 1951. The number of screenings performed on group members reached 25,000 in 1960 and jumped to 50,000 by 1970.[50] Screening programs for tuberculosis and then syphilis during the interwar and postwar decades appeared to observers to have become highly successful once cures became available. Looking for early-stage cancers had become a routine activity although screening was applied to restricted populations and chances for successful outcomes were considerably lower. By 1950, 250 cancer detection centers were functioning across the United States.[51] The PHS conducted the first community mass screening effort for diabetes from 1946 to 1949 in Oxford, Massachusetts. The goal was to estimate the prevalence of diabetes, evaluate diagnostic methods, discover cases and encourage patients to seek further treatment, and arouse awareness that testing was vital. In the 1950s, as blood glucose assessment methods became cheaper and automated, diabetes screening expanded.[52]

By the 1950s, multiphasic screening became the gold standard for proponents of screening. The dream was to save time and money by screening for many diseases simultaneously. Lester Breslow, chief of the California Chronic Disease Service, was a major advocate of this technique and claimed credit for inventing the term *multiphasic testing*.[53] In the late 1940s a demonstration test was designed in California "in the interest of economy"; it combined a health history, chest x-rays, blood specimens, and urine samples, to test for pulmonary or cardiac disease, syphilis, kidney disease, and diabetes, during a single screening process. The method was tested on 945 employees at 4 industrial corporations. A similar project was undertaken in 1948–49 in Massachusetts.[54] In July 1950 Surgeon-General Dr. Leonard A. Scheele announced that more than half a million people had been examined at multiple screening sites in Virginia, Georgia, Alabama, and North Carolina.[55] The President's Commission on the Health Needs of the Nation of 1952 (whose director of research was Breslow) recommended that "a bold attack be made on chronic disease with emphasis on multiple screening to detect disease early in physicians' offices, hospitals, industries, schools and health centers, promoted by the agency which has effectively pointed the way in preventing communicable diseases—the health department."[56] Although Breslow acknowledged the legacy of periodic health examinations, he did not have much faith in them, pointing out that they were time consuming, required far more physicians than were available, and were in any case unpopular among doctors.[57] These obstacles could be bypassed by combining the new, efficient tests now available into a single testing structure. Like physical examinations, testing programs were also meant to provide advice about health-enhancing behavior.[58]

Multiphasic programs proliferated widely during the 1950s and 1960s. One innovation was the development of mobile screening units in Oklahoma, Maryland, Arizona, and the District of Columbia.[59] A turning point was the development during the 1960s by Morris Collen of Kaiser Permanente of automated procedures that used computerized equipment and data analysis that speeded up the process.[60] Between 1969 and 1971, the number of multiphasic programs doubled to 140 nationwide. In 1973 it was estimated that some 3.5 million people would be screened during the year.[61] Screening could be aimed at risk factors as well as diseases. In 1974 newspapers reported on a $100 million federally funded testing project being carried out in 20 cities. The trial aimed to determine whether treating heart disease risk factors like hypertension, high cholesterol, and smoking could lower overall incidence of heart disease. Patients who fit the test criteria were asked to "return for a complete physical, including a blood test, a stress test, and a cardiogram."[62]

Despite the initial enthusiasm, multiphasic screening came eventually to be regarded as a failure. Although screening certainly uncovered many illnesses or abnormalities, public health officials and physicians began to question whether they actually had a positive health impact. Two major randomized controlled trials meant to evaluate the efficacy of multiphasic screening were initiated in the late 1960s. The first was performed by Collen and his colleagues at Kaiser

Permanente and involved 10,000 people with follow-up and data analyses at 7 and 16 years. The second was the South East London Screening Study Group, which surveyed 7,000 people with a 9-year follow-up. Neither trial found significant differences in these outcomes between study and control groups. Smaller trials produced similar results, as did many observational and retrospective studies. Added to the fact that false positive and false negative results were common, such trials doomed multiphasic screening.[63]

In the end what remained of the screening program was the not uncontroversial practice of screening for specific diseases, notably cancers. Screening for cervical cancer using the Pap test increased in the 1970s, with programs being developed by health departments in several states. The practice was hardly undisputed but nonetheless became widespread. Mammography was developed in the 1960s and began to be used as a screening instrument. It was not until the 1990s, however, that federal funds were earmarked for the establishment of cervical and breast cancer screening programs nationally.[64] By the last decade of the twentieth century, colorectal screening was also on the agenda, first through fecal occult blood testing and then, as techniques improved, through sigmoidoscopy and colonoscopy. Numerous other procedures were tested and practiced on a small scale.

Although it was subject to many of the same criticisms as multiphasic screening, periodic physical examinations seem nonetheless to have flourished. In 1979 the Canadian Task Force on the Periodic Health Examination found that preventive health measures "well supported by evidence" could be scheduled effectively during visits for acute and chronic care, rather than during periodic exams.[65] Despite such recommendations, regular criticisms, and little evidence of effectiveness, a population increasingly preoccupied with health status became progressively enamored of the practice. Unlike large-scale programs, it could be done in physicians' offices with minimal disruption of routines. Examinations also became shorter, simpler, and narrowly focused on finding illness, with a bit of multiphasic screening in the form of blood tests frequently included.[66]

Health Education and Promotion

Another activity that cut across specific diseases was health education. Like screening, this was not a new activity and was not restricted to chronic diseases. American life insurance companies and public health agencies had shown great faith in the practice early in the twentieth century. In 1943 the *Health Education Journal* began publication in Britain focusing primarily on infectious diseases. In the United States, the Society of Public Health Education was formed in 1950 and began publishing a journal, *Health Education & Behavior*, seven years later. The CCI placed considerable emphasis on preventive education. Nonetheless, its practice appears to have been somewhat overshadowed by screening and early detection during the 1950s and 1960s. A 1960 survey of Pap smear testing frequency conducted by the California Department of Health found that 85% of women had their test as part of routine physical examinations or at their physicians' urging; no mention is made of education to promote test-seeking.[67] There was, of course, growing interest in health education and in transforming personal habits, but during the 1960s, when Medicare and Medicaid increased access to the health system, a major goal of such education was making sure that people made use of healthcare resources, complied with medical guidelines when sick, and adapted to new roles when chronically ill.[68] While educational initiatives for cancer, diabetes, and respiratory and other chronic illnesses were certainly undertaken during this period, there is little evidence of coordinated programs for public chronic disease education.

As experiments in multiphasic screening gradually petered out during the 1970s, health education took on new importance, encouraged by the 1973 Report of the President's Committee on Health Education that led to the formation of the Office of Health Information and Health Promotion in the Department of Health, Education, and Welfare in 1976.[69] Some programs continued to focus on the management of sick people. The APHA published in 1975 a guide for healthcare workers "designed to assist the professional in planning for patient and family education and [that] may be used with any illness regardless of its etiology or chronicity."[70] Other programs focused on modifying behaviors, facilitated by the emerging notion of "risk factors" that permitted greater precision of action. Growing out of epidemiological studies, notably those dealing with the effects of cigarette smoking and the Framingham Study of Cardiovascular Disease, risk factors became a major part of the public health approach to prevention.[71] The initial goal was to identify high-risk populations that could be directed toward targeted screening. But during the 1970s and beyond, the elimination of risk factors themselves became the chief objective. Behaviors such as smoking, alcohol consumption, and sedentary lifestyle, and conditions such as hypertension, obesity, and high cholesterol, risk factors for many diseases, were made major targets of public health intervention. The National Institutes of Health in 1972 began a Multiple Risk Factor Intervention Trial to test the effect of various sorts of interventions to reduce risk factors for coronary disease.[72]

Prevention of risky behaviors or conditions was accompanied by a shift in health education practices. During the 1960s, public health educational work around chronic disease had taken a traditional approach—producing pamphlets and providing information about health resources, nutrition, and so on. Throughout the 1960s and 1970s there was an expanding body of research within the health education world on behavior modification, consumer participation, self-care, and the social norms that operated as barriers to healthy lifestyle choices. Specifically, there was a critical reevaluation of the effectiveness of health education methods and of the belief that "education" would necessarily produce significant changes in behavior. The anti-smoking campaigns of this era were a major arena for confronting the difficulties of modifying health behaviors.[73] Psychosocial models like the health belief model tried to analyze factors promoting or hindering compliance with preventive guidelines for the healthy and with treatment protocols for the sick. These made psychology a major disciplinary element in the chronic disease firmament that could be applied to many different diseases.[74] The Stanford Heart Disease Prevention Program was launched in 1972 with funding from the National Heart, Lung, and Blood Institute (NHLBI), and focused on media/educational interventions at the level of risk factors for cardiovascular disease. It became a major reference point for the promotion of risk factor surveillance and intervention. Around the same time the NHLBI launched a major public education campaign, the 1972 National High Blood Pressure Education Program, which included television and radio public service announcements, newspaper articles and press releases, and the organization of conferences and workshops.[75]

The 1979 publication of *Healthy People: The Surgeon-General's Report on Health Promotion and Disease Prevention* set the tone for public health in the coming decade that more than ever highlighted personal responsibility in maintaining health (although this was accompanied by a renewed interest in environmental risk factors).[76] Although health education provided information on such varied topics as vaccinations, maternal-infant care, and communicable diseases, chronic disease was a major component of this effort. To some degree this focused attention beyond single diseases, even if cardiovascular disease remained a prime target for risk factor intervention. Massachusetts began a coordinated statewide program directed toward multiple risk factors for cancer, heart disease, and cerebrovascular disease. Its Center for Health Promotion and Environmental Disease Prevention was devoted to screening and mass media campaigns in favor of diet, physical exercise and better nutrition, anti-smoking activities, and the elimination of environmental toxins.[77] On a national level, by the early 1980s, the American Hospital Association and the Centers for Disease Control each had centers for health promotion that were collaborating on a Health Education Project that produced more than a dozen reports, and that organized education activities in hospitals and health departments.[78] By then, educational activities had become professionalized with a newsletter, *Conference Calls*, published by the Conference of State and Territorial Directors of Health Education.

The emphasis on health promotion and individual behaviors did not go unopposed. In the 1980s there began to appear serious critiques of the health education strategy as at best an inadequate response to the social conditions causing disease and at worst "victim blaming." This critical view was inspired by homegrown American traditions of social reform and by developments in Europe, where health promotion was also becoming a dominant public health focus but where there existed a much stronger political left and legacy of social medicine. The World Health Organization (WHO) performed a synthetic function when it expanded the notion of health promotion to include interventions aimed at populations and social conditions.[79] While not without influence in the United States, this critical perspective remained a minority view; focus on individual behavior remained the dominant American approach to public health during the last decades of the twentieth century.

Problems of Organization

Despite the best efforts of the PHS, the autonomy of state and local health departments meant that coordination remained a problematic issue. Both the Regional Medical Program (RMP) created in 1965 and the Comprehensive Health Planning Program set up a year later were conceived as administrative answers to health coordination problems, but both came to be regarded as failures.[80] Of the two programs, the RMP is of particular interest since it was directly relevant to the chronic disease movement. Following from yet another President's Commission, it was intended to provide a national network of centers for heart disease, cancer, and stroke, and to coordinate research, public health, and medical service. Administered through the NIH, the RMP provided project grants to fifty-five different regions, primarily to voluntary agencies, hospitals, and universities rather than state and local health agencies. Although the RMP programming was categorical in focus, it may be seen as an attempt to devise a single comprehensive strategy that cut across the different chronic diseases.[81] In 1968 the RMP was transferred from the NIH to the newly formed Health Services and Mental Health Administration and combined with eight programs run by the National Center for Chronic Disease Control; all were placed in the newly created Regional Medical Program Service. By 1970, almost $30 million had been spent on the RMP, but its effectiveness as a program was already in question. The RMP lacked clear delineation of its role, and many of its activities overlapped with those of other programs. It was accused of spending as much as 40% of its operating costs on administration, a figure that was hotly contested by its defenders. In 1970 five of the nine chronic disease programs that had been integrated into the Regional Medical Program Service were phased out, and in 1976 the program was shut down entirely.[82]

If significant coordination was difficult in the American context, repeated attempts by the government to trim budgets of administrative agencies hardly helped matters. To be sure funds increased but their proportion of expenditures did not. In 1980 about $97 million, or 3.1% of the budgets of state health agencies, were devoted to chronic illness, whereas 38% was devoted to maternal and child health and 9% to mental health.[83] A decade later, the Association of State and Territorial Chronic Disease Program Directors found that during the fiscal year 1989, total reported expenditures for chronic disease control activities in the United States had risen to $245,371,377, but this amounted to less than 3% of expenditures in all surveyed public health agencies. Reported per capita expenditures varied widely, from $3.83 in California to zero in Oregon. Federal sources comprised only 20% of these funds.[84] Things improved somewhat during the early 1990s. In a later survey of 41 states, the association found that by 1994 chronic-disease-related expenditures had gone up by 22%, federal funds rose from 20% to 45% of all resources, while per capita expenditures rose from $1.05 to $1.21.[85]

If they could not coordinate local activities, national institutions could at least provide leadership and advice. The National Cancer Institute (NCI) continued to play such a role. In 1986 it initiated a major effort in "capacity building" for cancer control in local public health departments. The Cancer Control Technical Development in Health Agencies program committed $7.4 million to grants for activities like smoking cessation, diet modification, and cervical and breast cancer screening.[86] In the last decades of the twentieth century, the CDC became a major player in the chronic disease world. Founded during World War II as a branch of the PHS, the CDC in the late 1970s began taking a more concerted interest in chronic disease as part of a general expansion of its mandate beyond infectious diseases. It founded a Diabetes Division in 1977, collaborating with the health departments of ten states in implementing diabetes control demonstration projects. It reported in 1978 that cancers and cardiovascular diseases were among its top priorities and, like other branches of the PHS, provided local authorities and national associations with assistance and grants.[87] In 1984 it established the Behavioral Risk Factor Surveillance System with fifteen states participating in monthly data collection. The CDC had since the 1950s been running a household survey program to monitor the health status of selective populations. Under the new program, each state administered phone surveys (considerably cheaper than door-to door surveys) that gathered information on lifestyle and health-related behaviors. This became a major data source for risk factors and "health practices" used in the design and planning of state public health programming. By 1994, the program had expanded to all fifty states.[88]

The CDC in 1988 established the National Center for Chronic Disease Prevention and Health Promotion "to lead efforts that promote health and well-being through prevention and control of chronic disease." This meant preventing risk behaviors and promoting healthy behaviors, accelerating translation of scientific findings into community practices, and promoting social and environmental policies. In 1990 it sponsored a comprehensive breast cancer screening program through state public health departments. Within a few years it had ongoing programs for many major diseases.[89]

Together the concept of "risk factors" and what came to be known as "health promotion" helped advance the comprehensive view of *chronic disease* as a distinct administrative category. A good deal of risk factor epidemiology was about the identification of common underlying, predisposing, or predictive elements for multiple diseases. This allowed for the development of programming that targeted not the diseases themselves but the behaviors of otherwise healthy individuals. By 1982, the secretary of the Department of Health and Human Services could claim, "studies now show that most people can make daily decisions that influence their health and their vitality more than all of today's medicine."[90] Five years later, the report of the National Conference on Chronic Disease Control and Prevention concluded: "We must translate 'chronic diseases control'—a vague and uninspiring term—into a *coordinated and comprehensive lifestyle initiative* which will improve health and reduce the burden of illness for a *broad category of ailments*, while not losing sight of our objectives for individual diseases and risk factors."[91]

Conclusion

American chronic disease policy during this period has largely been judged a failure. This is true of authors like Daniel Fox who treat the issue comprehensively[92] and of those who examine specific issues: nursing homes, homecare, or public health policies. This view reflects the fact that among the many measures considered necessary, some were given priority because they reflected the interests and views of dominant groups within the chronic disease coalition and/or because structural conditions made some measures more difficult to implement than others. Some proposed solutions, like multiphasic screening or making general hospitals centers of chronic care, turned out to be ineffective or excessively expensive. Rehabilitation has proven to be far from useless but it has not been a panacea. Health promotion has had positive effects (and has changed life for many) but clearly does not influence segments of the population that need it most or address deep socio-structural issues. And the fact that so many Americans have lacked adequate health insurance and consequently satisfactory healthcare has negatively affected both outcomes and our perceptions of them. Overall, mixed successes and failures resulted inevitably from deeper causes: the elaboration of a hugely ambitious, overoptimistic, and vague program by a wide-ranging coalition of interests that lumped together wildly dissimilar problems; decentralized centers of power that allowed for much local innovation but often proved unable to solve problems of wider coordination; and, not least, the extraordinary complexity of problems being confronted, which even now do not lend themselves to easy solutions. With so much involved in the response to a chronic disease "problem" that appeared to grew daily as the population aged, as death was increasingly evaded at the cost of leaving behind morbidity and infirmity, as more and more risk factors and conditions were transformed into chronic illnesses, and, perhaps most significantly, as our expectations of health rose exponentially, it was inevitable that many initiatives would be perceived as failures.

On the other hand, there is no question that the chronic disease movement has transformed our understanding of health and illness. We all now talk the talk of risk factors, health promotion, screening, and surveillance, even when we are being critical. Not everything we do or disparage is the result of the chronic disease movement (the proliferation of childhood vaccinations, monitoring of pregnancies, annual flu shots) but there is no question that the attention given to chronic disease has contributed mightily to the way we approach health and illness. The result of all the American agitation around chronic disease was a compromise or, more correctly, a series of compromises. But it was not compromise between old and outdated tactics conceived during the era of infectious diseases and new, more appropriate responses. It was rather a compromise among many interest groups with divergent goals—bringing chronic disease into mainstream biomedicine, defending private medical practice, expanding and to some degree medicalizing public health, keeping a lid on expenses. It was a compromise among various strategies for dealing with chronic diseases—biological research, nursing homes, and screening. Finally, it was a compromise between somewhat theoretical strategies for coping with new problems and the complex realities of American political and social life.

PART II / Chronic Disease in the United Kingdom and France

CHAPTER EIGHT

Health, Wealth, and the State

For much of the twentieth century, the American preoccupation with chronic disease was exceptional. Most of the issues associated with such illness were not specific to the United States but they were perceived, understood, and classified in different ways in other countries. I illustrate this point in the following two chapters by examining developments in the United Kingdom and France. My goal is to (1) underscore the uniqueness of the American focus on "chronic disease"; (2) attempt to understand how distinctive national realities produced different conceptual frameworks and axes of disagreement that structured debates about healthcare policy; and (3) explore the international circulation of the notion "chronic disease" during the second half of the twentieth century and beyond.

Before examining each nation in detail, a number of general observations are in order. Britain and especially France were during the first half of the twentieth century older societies than the United States. In 1921 around 7.5% of the population was aged 60 and older in the United States, whereas the comparable figures were 13.7% in France and 9.4% in England and Wales. In 1950 those 65 or older made up 8.3% of the American population and roughly 11% in the two European nations. Such discrepancies continued well into the 1970s.[1] In France population aging was a longstanding phenomenon due to traditionally low birth rates. In both European countries, the deaths of large numbers of young men during the two World Wars raised the overall population age. Finally, neither European country experienced prolonged high rates of immigration characteristic of the United States, that land of such extravagant promise. This reality had paradoxical results. Since chronic illness was associated with age, both Britain and France should have experienced a significantly more intense chronic disease problem than did the United States. This did not occur, however, because the sad reality was that no one was particularly concerned with older persons. American preoccupation with chronic illness was predicated on the assumption that it affected large numbers of young or middle-aged men and women. This made their cure and rehabilitation a significant national priority with major implications for the future of the nation. The illnesses of older people, in contrast, appeared to be part of the natural order and had little if any consequences for the productivity or might of the nation. Precisely because chronic disease was identified with the elderly population, it was largely ignored in the United Kingdom and France during the first half of the twentieth century. The poorest members of this group were placed in national networks of institutions: Poor Law or successor institutions in Britain and hospices in France. When their condition did emerge as a major humanitarian issue during the second half of the century, the chronic diseases that afflicted many were perceived as an aspect of service provision rather than one of health status.

A second major difference had to do with national wealth. The United States emerged in the twentieth century as the richest nation in the world and as a superpower. This position was enhanced after World War II when it became the undisputed leader of the noncommunist world. It had more money than any other country to spend on healthcare at a time when health was becoming a major consumer demand. Without a national health insurance system to fund, postwar American authorities had greater financial margin and, I have argued, political incentive to confront challenges like chronic illness that threatened few entrenched interests. France and Britain emerged from the war relatively impoverished and with major reconstruction tasks to undertake. They had to renovate and modernize aging and technologically outdated hospital systems that now served larger patient populations and that utilized more expensive equipment and techniques. Like most nations, they were dealing with specific illnesses like tuberculosis, venereal disease, and increasingly cancer—some of which might be considered chronic; but no one was seeking to confront a huge and amorphous new category like "chronic disease" whose costs were impossible to calculate. The political choice to invest large amounts of money in a national healthcare service or health insurance scheme precluded investment in other areas but also defined the issues to be faced during the following decades. These included pressure to extend coverage, increase equality of access, manage relations with powerful medical professions, and, not least, control costs.

The growing wealth and power of the American republic combined with the relative paucity of centuries-old entrenched institutions like the Poor Law system in the United Kingdom or hospices in France had consequences for the way problems were perceived. One cannot read American medical and public health literature published before 1970 or so without noting the profound optimism about the capacity of Americans to solve problems and in particular to cure and prevent diseases given enough science, money, and American knowhow. Europeans also had faith in science, but their sense of optimism was more subdued following two devastating World Wars and the subsequent loss of colonial empires. The age structure of the population was also relevant. It was easier to be optimistic about problems of the young than those of the old, and facing a less crushing burden of care for elderly people may well have created the space necessary for Americans to think of chronic disease as remediable.

That being said, it was impossible for Europeans to ignore what was going on in the United States. Medicine, like science, was an international activity. Medical journals in all countries had always taken an active interest in foreign developments. By the mid-twentieth century, American medicine represented the standard of excellence by which all developed countries measured themselves. European researchers and clinicians traveled to the great medical centers of the United States. European physicians met American colleagues at international meetings, as did eventually epidemiologists and health planners. Finally, if Europe did not face exactly the same problems as the United States, it faced similar ones, including the presence in hospitals of large numbers of long-term patients who did not seem to benefit from the expensive skills and technologies that were located there. As early as 1951, the International Congress on Hospitals in Brussels was devoted to chronic and elderly patients. That this was the subject of an international meeting undoubtedly reflected the fact that hospitals everywhere were seeking to reconcile their increasingly acute orientation with the realities of their patient populations. But American influence was also clear. The well-known American chronic disease activist E.M Bluestone, former director of the Montefiore Hospital in New York City and famous for pioneering hospital-administered homecare, delivered the keynote address at this meeting.[2] Such influences did not just move in one direction. British figures played a major role in developing the risk factor epidemiology on which much chronic disease policy was based. There were domains like geriatrics where European nations were far ahead of Americans. In 1969 the US Senate held hearings on the economics of aging and invited leading European gerontologists to present reports.[3] And perhaps the single most influential political document on the need to revitalize healthcare systems to take account of the changed nature of disease (the word *chronic* appears repeatedly) was written by an erudite *Canadian* minister of health.[4]

Another more focused mechanism for transmitting American ideas or at least getting Europeans to engage with them was the World Health Organization (WHO). In 1957 the European Section of the WHO held a symposium in Copenhagen on the Public Health Aspects of Chronic Disease.[5] Public health administrators from thirteen countries attended, as did some experts in chronic diseases. One of the lectures was by Lester Breslow (read by a third party), now a leading figure in the American chronic illness movement. The goal of the meeting was "a preliminary evaluation of the magnitude of the problem of chronic disease" in Europe, discussion of its various aspects, particularly those relating to public health, and of possible future action by nation states and the WHO. The focus was on the age group 40–64, and on four disease groups: malignant neoplasms, diabetes, cardiovascular disease, and rheumatic disease. Subsequent meetings and technical committees on the role of hospitals and ambulatory and domiciliary care included American figures like Bluestone and spread their ideas about the need for hospitals to play a greater role in prevention, rehabilitation, and domestic care.[6] An expert WHO group on rehabilitation advocated greater integration of rehabilitation into medical services, as well as expansion beyond joint and limb work to chronic internal diseases.[7] The organization commissioned the first guide to screening practices that set strict criteria for introducing programs that eventually influenced American policies. Similarly, American public health figures were influenced by the WHO's definitions of positive health and health promotion.[8]

Initially, much of the work of WHO committees involved stating fairly general principles or organizing meetings where representatives of different countries could discuss their activities and experiences. This meant that influence was rarely direct; WHO pronouncements frequently provided frameworks or arguments that could be used by local actors pursuing diverging, even conflicting, agendas.[9] More practical projects were disease specific. At the request of member states the organization in 1968 began a cardiovascular control program that included multiple national and professional partners (like the International Society of Cardiology). The first phase covering 1968–72 focused on ischemic heart disease and included projects to collect information, and to experiment with prevention, treatment, rehabilitation, and long-term follow-up of patients. In a second phase from 1973 to 1977, the program was expanded to include hypertension, stroke, rheumatic fever and rheumatic heart disease, congenital heart malformations, and chronic chest diseases.[10] This reflected several realities in Europe, including the fact that the more general notion of chronic disease or illness was understood less as a coherent policy concept than as a loose collection of illness problems. This was the case until the first decade of the twenty-first century, when the chronic disease activities of the WHO intensified, while British and French policy planning began to take this category seriously as a way of introducing disease management approaches to healthcare.

CHAPTER NINE

Alternative Paths in the United Kingdom

M any of the same conditions that promoted concern with chronic illness in the United States also existed in the United Kingdom. During the interwar period, there was considerable talk about disease prevention by major figures like Sir George Newman, longtime chief medical officer in the Health Ministry, and Lord Dawson of Penn, president of the Royal College of Physicians.[1] Public health officials were agitating to extend their functions beyond infectious diseases to include curative medicine. Public awareness of cancer was at least as great as in the United States; the Imperial Cancer Research Fund predated the precursor of the American Cancer Association by a decade and in 1922 the Ministry of Health established an ongoing Departmental Committee on Cancer, which in its first five years published fifteen reports.[2] In 1934 the Royal College of Physicians established the Committee on Chronic Rheumatic Diseases, transformed into the Empire Rheumatism Council two years later.[3] There was widespread awareness that many people were not getting the care they needed because institutions were inadequate. There was overlap and competition among different agencies at the local level. And sick older people were not a priority for anyone.

And yet, despite the many similarities, the situation in the United Kingdom differed substantially; during the interwar years, one finds almost no reference to a "chronic disease problem." The word *chronic* remained an adjective that could apply to any condition that lasted a long time. When used a noun, *chronics* had the usual ambiguity. It might refer to the lingering sick requiring no nursing or medical supervision, the incurable and incapacitated needing institutionalization, or those with any longstanding condition.[4] The situation changed during the post–World War II period when chronic disease did become a social problem, but one that referred almost exclusively to the sick elderly population. This understanding of chronicity dominated British medical and policy thinking for several decades and pulled it in directions diverging from those followed in the United States. Only during the 1960s did the American meaning of "chronic disease" come into general circulation in Britain. And while its use was certainly consequential, chronic illness understood in this way did not become a major political issue as it had in the United States.

The Interwar Years

One critical difference between the two nations had to do with almshouses or poorhouses and the infirmaries and hospitals associated with then. Whereas such American institutions were local and gradually emptying out as their inmates were moved to more specialized establishments, British Poor Law medical institutions were part of a national public system (locally administered) that constituted about 60% of the nation's hospital beds in the 1920s, and that was far larger than either the voluntary (private-nonprofit) or public municipal hospital sectors. That these institutions were inappropriate for dealing with illness was clear to many by the early twentieth century; but so many institutions could not simply be abandoned and the sector was too large to be dealt with by anything other than national legislation. The famous Poor Law Commission of 1905 was clear about the need to abolish the Poor Laws but divided between a majority and a minority regarding the type of service that should replace it. Consequently, nothing was done about this issue until the end of the 1920s.[5]

Meanwhile, the British state was instituting other welfare measures. The 1908 Old Age Pensions Act provided for a small weekly pension for people over 70 years of age that allowed at least some of the elderly to avoid entering Poor Law institutions. The National Insurance Act of 1911 established health and unemployment insurance to be paid for through contributions by the state, employers, and employees. Only part of the population was covered and hospital care was not included, but the measure further reduced the potential population that might end up in poor houses due to loss of work from illness. Together with the rise of nongovernmental hospital insurance, the Insurance Act also generated growing pressure to properly finance basic medical services and extend coverage.[6] These uncompleted reforms rather than "chronic illness" were the chief issues receiving attention in interwar Britain. Other laws provided for free treatment of uninsured tuberculosis patients and a school medical service. In 1919 a Ministry of Health was created to centralize many health-related functions. The Public Health (Tuberculosis) Act of 1921 required county and county borough councils to provide sanatoria care and after-care services for tuberculosis patients, and to remove highly infectious patients to hospitals. But the reform that was most pertinent for the issue of chronic illness was the Local Government Act of 1929, which along with the 1930 Poor Law Act and several subsequent laws, disassembled the administrative structure of the Poor Law and placed it under the supervision of local authorities.[7]

One of the chief aims of the law was to remove pauperism as a criterion for access to medical treatment. Another was to unify or at least coordinate health services provided by local authorities. Local councils were to create public assistance committees (PACs) to continue welfare work but were encouraged to transfer health services to a public health committee (PHC) that was also responsible for municipal public hospitals. This meant that Poor Law hospitals and workhouses could be appropriated and managed by PHCs, allowing for planning and coordination within municipal systems. As a result, nearly fifteen thousand hospital beds were added to the public municipal sector by 1938. Among the many factors that determined local willingness to transfer institutions were the quality of the institutions and buildings that were made available and their appropriateness as public general hospitals.[8]

The consequences of the act were, it is generally conceded by historians, highly uneven and dependent on local conditions. To the extent that Poor Law institutions were taken over by local PHCs, they added to the nation's stock of general acute-care hospitals that in some cases at least provided chronic care as well. For the most part, however, no one was much concerned by infirm older people, considered incurable and not essential to the future of the nation. To the extent that less well-endowed institutions remained under the jurisdiction of PACs, it was predominantly the "chronic and elderly sick . . . always the residuum of the Poor Law services" that filled them and thus "had a much greater risk of . . . receiving second-class care."[9] Surveys by the Ministry of Health found that some former Poor Law facilities were of very low quality with gross overcrowding the norm and a relatively poor level of provision for indigent sick patients.[10] But this situation did not provoke undue concern. The accepted wisdom during this period was that the country needed modern hospital medicine rather than low-level chronic care. Central government subsidies also encouraged local spending on the health of children and young adults. Old and chronically sick Britons were not an appealing target for investment.[11]

Chronic disease remained relatively unproblematic during the interwar years and was generally understood as a problem of old age. Tuberculosis was, of course, a major issue but was not usually considered a chronic disease because of its contagiousness. There was considerable interest in cancer and to a slightly lesser degree in rheumatism/arthritis, evident in medical journals and the *Annual Reports of the Chief Medical Officer of the Ministry of Health*, but there was little interest in classifying these diseases as part of a larger category or bigger problem. One major factor was that Britain was an older society than the United States (see chapter 8). Public concern about sick elderly people was not pressing; they were hard to cure, unsatisfying to care for, frequently sick for lengthy periods, and unlikely to become economically productive. The existence of a large elderly infirm population thus made it hard for British reformers to argue, as Americans did, that chronic illnesses were to a considerable degree a problem of young and middle-aged adults, and that confronting the problem was an issue of long-term national "vitality." The politics of institutional appropriation that segregated many elderly people in residual Poor Law institutions heightened the identification of chronic illness with them, even as it made these individuals largely invisible to the general public. It was widely believed that limited resources could be used more effectively on younger, acutely ill patients. The centuries of stigma attached to the Poor Laws certainly played a part as well. The bottom line perhaps is that there were few if any groups in the United Kingdom with strong motivations to expose or transform the plight of "chronics."

As in the United States, British statistics seemed to show a dramatic rise in mortality from cancers and coronary diseases. But statisticians and leading health administrators were generally skeptical about this data, arguing repeatedly that it was the result of an aging population and better diagnostic methods. The skepticism was to some degree validated when the Cancer Commission of the Health Organization of the League of Nations in 1927 criticized the "unsatisfactory nature of certification of causes of death" and urged countries that collected official statistics to improve their systems. British statisticians responded by introducing a "standardized mortality rate" meant to eliminate effects of changing factors like age and sex; this procedure significantly reduced cancer mortality rates, although even figures calculated in this way suggested that cancers of certain organs, notably the lungs, seemed to be rising somewhat among men.[12]

Another major difference had to do with political context. American reformers in the 1930s were advocating a dramatic restructuring of healthcare. In pursuit of this goal, they sought to demonstrate that the health situation in the nation was terrible and damaging to national prosperity. They stumbled on chronic illness as a way of making this case, but one can easily imagine other issues serving similar purposes. The British, in contrast, had by now taken significant steps toward gradualist reform of healthcare. The principle of health insurance was accepted and pressure was building to extend it to a larger proportion of the population and to cover hospital services; the Poor Laws had been abolished and most Poor Law institutions seemed slowly and unevenly on the way to being transformed from caretakers of the poor to healers of the sick. The state had already assumed some responsibility for cancer, regulating the purchase and distribution of radium to hospitals and setting up a system of cancer centers associated with major teaching hospitals.[13] There was little reason to use poor health conditions as a rationale for change since change was underway. There were, if anything, excellent political reasons for reformers to emphasize successes rather than failures in order to maintain momentum for continuing an ongoing reform process.

One of the forces contributing to American interest in chronic illness was a public health movement seeking to expand beyond infectious diseases into curative medicine. In both Britain and the United States, the process was already underway in such areas as maternity, child health, tuberculosis, and venereal disease services but ambitions were far greater. The Local Government Act satisfied many of these wider aspirations by granting local public health committees, within which medical officers of health (MOsH) played a key role, overall responsibility for coordinating municipal hospital facilities and appropriating Poor Law institutions. This was a deliberate attempt to unite public health prevention with communal healthcare and to promote "wider application of the principles of Preventive Medicine." The MOH was now said to be "in charge of the health and complete medical services of the whole community."[14] Public health workers were if anything overwhelmed by the resulting administrative work in the healthcare sector. Historians of public health have recently and effectively refuted the claim that this shift led to the abandonment of traditional preventive and sanitary activities.[15] But what is unquestionably true is that between older functions that were retained, new administrative functions, and the health education programs that were then getting started, MOsH were amply occupied. There was little pressure from that quarter to further expand the domain of public health.

Postwar Concern with Older People

Largely ignored during the interwar period, the condition of elderly hospital patients burst onto the public stage after World War II. A key trigger was a series of regional reports based on a survey of hospital accommodations carried out during the war by the Nuffield Provincial Hospitals' Trust and the Ministry of Health that was published in 1945–46. These reports gave the impression of inadequacy largely across the institutional board. Nationally about 35% of all beds were in voluntary hospitals and 50% in the now much expanded municipal hospital sector. But whereas those who could be actively treated were thought to enjoy improved hospital access, chronic patients had with some significant exceptions been parked in the inferior facilities of welfare authorities. According to Charles Webster, about sixty thousand elderly sick Britons were in 1939 in public assistance institutions (former workhouses). These were overcrowded, unsanitary, and inefficient. About 15% of total hospital beds were in such institutions (the figure was more than 50% in some boroughs). Half of all chronic patients were in the public assistance sector. In twenty-one out of thirty-six boroughs for which figures were available, 100% of chronic beds were in the public assistance sector. Not surprisingly, these seriously understaffed institutions were seen as a dumping ground for indigent elderly and chronically ill people. The war had made things even worse, as many of these patients were moved by the Emergency Health Service from cities into outlying areas and kept bed ridden.[16]

The Nuffield surveys were only one aspect of increasing concern with an expanding older population. In 1959 the sociologist Peter Townsend pointed out that interwar social surveys had focused on poverty and unemployment. This attention had been superseded after the war by surveys about elderly people; he discussed thirty-three surveys on this topic published between 1945 and 1958. Their quality was not particularly good, he suggested, and they dealt with numerous issues, including social services, household management, and adjustment to retirement. But health issues were part of the picture especially in the case of about 10% of elderly persons who were housebound because of infirmity.[17] Some of those who worked with this population began agitating for programs to deliver adequate medical care. The work of Marjory Warren in the 1930s and 1940s suggested that intense rehabilitation could improve the condition of many and get them out of hospitals relatively quickly. A movement to set up geriatric hospital wards thus came into being. A number of physicians including Lord Amulree, Lionel Cosin, J. H. Sheldon, and T. H. Howell, also advocated for change and developed new rehabilitation techniques. Their views gained traction due to the surveys that showed that so many elderly patients received inadequate care or no care at all. Others believed that many precious acute hospital beds were wasted on elderly "chronics." Doctors, hospital administrators, and politicians came to agree about the need to get "bed-blockers" out of acute-care hospital beds and into more appropriate settings.[18]

The creation of the postwar National Health Service (NHS) and of the nationalized hospital service promised to benefit elderly Britons who were heavy users of medical resources. It was hoped that care could be improved, rationalized, and calibrated to the degree and kind of disability manifested.[19] But this hope, it is now widely argued, was largely dashed.[20] By 1949, the British Medical Association (BMA) and other agencies were reporting that general hospitals were refusing to admit many older patients. There were those in the hospital sector who would have liked to offload elderly "bed-blockers" entirely onto local welfare authorities but the latter predictably resisted. A less controversial alternative was to create geriatric services focusing on early treatment and rehabilitation with the goal of shortening hospital stays. By 1947, the BMA supported this vision of geriatrics, which soon became the official positions of doctors working with older patients.[21] Linked with this was considerable support for homecare. Together geriatrics and homecare, it was hoped, could relieve pressure on hospitals.

The NHS, however, was slow to implement these ideas. Contemporaries blamed difficult economic conditions. Historians have variously blamed failure of government leadership, little concern with unproductive segments of the population, opposition of hospital consultants to investment in geriatric wards, lack of enthusiasm by local authorities for developing homecare, and the stigma and institutional logic that remained associated with Poor Law institutions. Once action was possible, historians argue that support for geriatric medicine was mobilized not to improve standards of care but rather to justify severe restrictions on access to long-term hospital care for the elderly.[22]

The Ministry of Health only began to play a significant role in these matters in 1953, when it set up a committee to examine services to the chronically sick and elderly which became known as the Boucher Committee after its chairman, C. A. Boucher. The report, published in 1957, concluded that the level of hospital provision was generally sufficient and that "any continuing problems were due to the inefficient use and distribution of beds." The solution was thus not more beds but good rehabilitation services, adequate welfare accommodations, and better health services (especially homecare). As a result, the Ministry sent out a circular accompanying the report, which suggested that everything possible be done to keep old people at home. It proposed, in fact, that regional hospital boards limit provision for the chronic sick to 1.2 beds per 1,000 population, a guideline that, when respected by local authorities, represented a significant reduction in the number of hospital beds available to elderly "chronics." The Ministry, however, did not invest much in either rehabilitation or homecare and limited itself to "exhorting" local authorities to do so. "The overwhelming priority appears to have been to limit the 'burden' being placed on the hospitals."[23]

Costs were, of course, an important factor. The government did not have enough money to construct new hospital facilities until a building program was approved in 1962. But even new construction reinforced and extended the policy of bed norms. In 1963 the Local Health and Welfare Plan was passed with the aim of increasing domiciliary services. But many local authorities preferred instead to expand residential services. As a result, the number of residential homes increased substantially during the 1960s. At the same time, the freeze on hospital beds for old people combined with the increased size of the aged population exacerbated the shortage of beds. The average length of stay in long-term hospitals increased and waiting lists grew longer.[24] A study of nursing care in hospital wards from 1955 to 1980 concludes that despite the dedication of many nurses, "the care received by the elderly patients was often uncaring, routine and regimented."[25] Nonetheless, as many commentators have noted, support of the Ministry of Health was critical in allowing geriatrics to become established in spite of significant opposition from parts of the medical profession. The number of geriatric consultants rose from 98 in 1963 to 214 in 1971. The first university chair in the field was established in 1965. The status of geriatricians remained low within the profession and their appointment did not necessarily lead to effective services. As late as 1978, forty-two health districts in England lacked any geriatric beds.[26]

Geriatrics was geared to short-term rehabilitative care, leaving longer-term care in old Poor Law institutions until the 1970s, when most were replaced. (New facilities, however, remained under the control of welfare rather than health authorities.) The treatment of elderly patients became a serious public issue during the 1960s, with the publication of numerous books and articles, most notably Peter Townsend's *The Last Refuge*, which exposed the inadequacies of residential institutions, and Barbara Robb's *Sans Everything*, which more impressionistically exposed the cruelty and incompetence of much elderly care under the NHS.[27] In response, successive governments regularly spoke about and sometimes tried to direct more funds toward the hospital care of elderly "chronics." They were not very successful, it is claimed, because hospital consultants with other priorities dominated decision making at the local level, and because measures introduced in the 1970s to increase funding for elderly care were followed by an economic crisis that made implementation impossible. During the 1980s, a large influx of social security funds supported private residential care, relieving some of the pressure on hospitals.[28] But this money soon dried up. Meanwhile, the NHS was significantly reducing the supply of hospital beds. Between 1969 and 1999, the number of acute beds fell by 35% but the number of chronic beds was reduced nearly by half.[29] The British government, in Jane Lewis's view, was consistently pushing the boundary between NHS medical care, which was free and paid for by taxes, and social care, for which users paid according to means tests, firmly in the direction of more social care and greater client charges.[30] More geriatric units and more and better homecare remained the solution of choice for dealing with sick elderly people. But not everyone accepted this reasoning.

Thomas McKeown and the Birmingham Surveys

During the 1950s, several cities in Britain were sites of active efforts to improve the condition of older people.[31] One of these was Birmingham. In the late 1940s, Sir Arthur Thomson, consultant physician at the Birmingham General Hospital, vice principal of the local university, and chairman of the Birmingham Regional Hospital Board, set events in motion by directing a survey of the records of the largest chronic-care hospital in the city, the Western Road Infirmary. Results were presented in two lectures and then published. Thomson's approach was descriptive and a cursory effort was made to classify patients. He was not aggressively critical of the hospital, but did mention inadequate preadmission examinations of patients, poor facilities, and shortage of nurses. He suggested that the elderly should be treated and studied scientifically, especially in teaching hospitals, so that medical students received training in their care. He did not, however, favor the development of geriatric medicine as a specialty, wishing to keep the care of elderly patients within general medicine. Nonetheless, his goals were similar to those of geriatricians: speedy care during short hospital stays of less than three months would produce more recoveries and fewer patients with hopeless infirmities due to long inactivity and lack of treatment.[32]

Shortly thereafter, the surveys were taken over by Thomas McKeown (soon to become modestly famous) and Charles Lowe, respectively professor and lecturer in social medicine in the Birmingham University Department of Public Health. The goal shifted to identifying the proportion of chronically sick (predominantly elderly) persons requiring skilled medical or nursing care in hospitals and those who could be better and more cheaply cared for elsewhere. Over the next few years, the two surveyed a broad range of institutions and analyzed patient populations.[33] After publishing several descriptive articles, they wrote a policy paper in 1952. Their aim—rational distribution of patients—was common to a number of physicians then dealing with older people in institutional contexts. Marjory Warren, for instance, was at the West Middlesex County Hospital in the 1930s when the appropriation of a Poor Law institution brought hundreds of older chronic patients to her hospital. She divided up her charges among five groups defined primarily by physical and mental capacity, ranging from the more or less ambulatory to those suffering from senile dementia. She believed strongly that all should be in general hospitals where adequate medical care was available and where they would be part of the medical studies curriculum.[34] William Hughes developed different categories for chronically ill patients at the Stapleton Geriatric Center in Bristol. Certain categories like "mental defectives" and psychotics belonged in other institutions, as did the ambulatory. The acutely ill clearly belonged in the Geriatric Centre but what he called the "frail ambulant " and the "bedfast" who did not require skilled medical or nursing care but who were unable to care for themselves posed a dilemma. The long-term goal was to develop smaller "peripheral" hospitals into long-stay centers for such patients but Hughes admitted in 1952 that no practical scheme for the aged sick currently existed.[35]

McKeown and Lowe were thus working within this tradition of institutional classification. They defined four categories of patients and the kinds of institutions that they needed. Group 1 consisted of patients requiring frequent medical attention or skilled nursing care and who belonged in special geriatric wards of general hospitals. Their reasons were practical; "the care and subsequent disposal of these patients require the attention of staff with a special interest in the management of old people."[36] They nonetheless believed that geriatrics should be integrated within general medicine rather than exist as a separate specialty. Group 2 was composed of patients who required relatively little medical or skilled nursing care but were confined to bed and thus required simple nursing care like washing, feeding, and lifting. The authors suggested that these patients be cared for in long-term annexes of hospitals, preferably on hospital grounds. The reasons for keeping them under the auspices of hospitals had to do with the distaste doctors and nurses showed for such patients. They could not be asked to care fulltime for such patients in isolated institutions. But if this was part of their normal work "there seems no reason why they should not accept it as an occasional obligation under a rotational scheme."[37]

Neither Group 3, ambulant patients needing simple nursing and occasional medical supervision, nor Group 4, ambulant elderly persons with no need of medical or nursing care, belonged in hospitals. If they could not live at home, they were ordinarily admitted to welfare homes or hostels administered by local authorities, a practice that the authors endorsed. The welfare and hospital administrations should cooperate in the provision of homecare, which was always preferable to institutional care. Such changes the authors admitted could only be implemented over a long period. But they proposed their views as a blueprint for future developments. In sharp contrast to their American counterparts, as well as British geriatricians, they said little and seemed rather pessimistic about rehabilitation.[38]

In further work over the next decade, McKeown and colleagues continued their surveys and applied their schema to mentally ill and tuberculosis patients. They found that a significant proportion of these patients, like those in chronic-care hospitals, could have been discharged if proper housing and homecare were available. In 1959 the geriatrician Joseph Sheldon did a survey of chronic-care hospitals in the region that was critical of many existing facilities. Nonetheless, by 1966, Birmingham had nineteen consultant geriatricians and improved accommodations for geriatric units. Meanwhile, McKeown and his colleagues continued their work and expanded their vision. McKeown was not, like another pioneer of social medicine, John Ryle, a "medical humanist" seeking through the social sciences a synthetic view of medicine.[39] He was rather a moralist, deeply troubled by the exclusion of certain groups from the most up-to-date medicine epitomized by the acute-care hospital. The elderly sick were the first such group that he dealt with and they remained a major concern. In an article published in 1961, he and his collaborators found the distribution of patients among different institutions was largely a function of age: regardless of the diseases they suffered from, young people were admitted to general hospitals while elderly people were usually sent to a chronic-care hospital where they frequently did not receive appropriate care. Despite all the reforms introduced by the NHS, they dryly concluded, "there has in fact been remarkably little change. The main difference is that whereas the common feature of patients in chronic-care hospitals was formerly their destitution, it is now their advanced age."[40]

McKeown came to realize that other groups were also excluded from general hospitals and mainstream medicine. The most numerous were sufferers of mental illness, segregated like elderly "chronics" in special, largely inadequate institutions. Following a report by the Ministry of Health in 1956 on planning hospital services for the mentally ill, the Birmingham group surveyed this institutionalized population. Classifying "patients according to their medical nursing and social needs," they found that about three-quarters needed only simple nursing care or personal supervision and about 12% could be discharged if they had a suitable home. Only 13% required full hospital care. Patients could thus be divided among appropriate units based on these categories of care.[41] During the 1960s, McKeown's group would discover yet another marginal and excluded group: the intellectually "subnormal." Thus, without resorting to the American concept of chronic illness, McKeown developed a broader vision of multiple but linked populations, defined not by disease categories but by exclusion from general hospitals and by inadequate care.

In 1958 McKeown published one of his most important policy-oriented papers on the organization of hospitals.[42] Here he criticized the separation of hospitals for the chronically (by which he meant elderly) and mentally ill from general hospitals, a separation that he attributed to historical contingencies that were no longer applicable. As in his 1952 article, he proposed a division based on the intensity and nature of patient needs but added a new wrinkle. Rather than dividing patients among inclusive general hospitals and remote segregated institutions, he proposed that his various functionally based institutions be housed in a series of smaller buildings within a common hospital complex—what he called "a balanced hospital community." This would make moving patients into the appropriate institution more practicable but his stronger argument was that this would lead to a more equitable distribution of doctors and nurses. Since no one wanted to work exclusively with elderly and insane patients, the only way to ensure adequate care was to have the same staff serving all populations. Medical and nursing students would see a hospital population that was representative of existing illnesses and that they would recognize as the responsibility of medicine. Yet another advantage of this configuration was to stimulate research on neglected conditions.[43] Realizing that both financial constraints and entrenched institutional interests would make implementation difficult, McKeown saw the hospital community as a model for institutions that might be built in the future.

McKeown returned to this idea again and again during the 1960s. A demonstration project in Birmingham was organized but it is not clear that this led to permanent changes. It is certain that McKeown's ideas were not generally implemented in the United Kingdom although it did become more common to create psychiatric units within general hospitals, addressing at least one of his concerns. In 1976 the Nuffield Foundation financed the creation of a day hospital in Sheffield for patients who had pre-terminal cancer and chronic disease.[44] While his attitude toward rehabilitation became more positive, McKeown remained convinced that emphasis on acute care was not justified by either historical or current mortality and morbidity patterns. Although his controversial thesis that the historical decline of mortality was due primarily to improved nutrition and living conditions has been debated to death, only one commentator, Ronald Green (to my knowledge, never once cited, although he is listed in a single bibliography), has linked this thesis to McKeown's wider commitment "to direct our attention to those substantial minorities whose care and treatment have often been forgotten in the quest for more sophisticated medical achievements." If McKeown downplayed scientific medical treatment it was "because he wants us to reach out to classes of persons, including the mentally subnormal and handicapped, the mentally ill and the aged sick, whose medical needs, he believes, have been insufficiently attended to in the past."[45]

McKeown was one of two prominent British health thinkers of that era to move beyond a notion of chronic illness as a problem of old age. He did not propose a radical departure from this traditional view but a modest extension to include several other categories of patients unfairly excluded from the healthcare system. And his central concern, despite all the historical epidemiological statistics he collected and mobilized, was not disease but proper care. A second major thinker, Jerry Morris, took a very different and more radical approach.

Jerry Morris, Chronic Disease, and the New Public Health

By the 1960s, Thomas McKeown had accepted two key ideas of the American chronic disease movement: developed societies faced a new set of illnesses and conditions that could not be dealt with by traditional means; and one could best prevent such diseases by modifying individual behaviors, in addition to more traditional strategies like eliminating environmental risks. But McKeown was a follower rather than a leader in this respect. Acceptance of chronic illness as a new public health problem was gradually spreading in Britain, but without the sense of crisis felt in the United States. John Ryle, the first professor of social medicine in Britain, sounded much like American colleagues when he wrote in 1948 that public health "has been largely preoccupied with the communicable diseases, their causes, distribution, and prevention. Social medicine is concerned with all diseases of prevalence, including rheumatic heart disease, peptic ulcer, the chronic rheumatic diseases, cardiovascular disease, cancer, the psychoneuroses, and accidental injuries—which also have their epidemiologies and their correlations with social and occupational conditions and must eventually be considered to be in greater or less degree preventable."[46] Ryle was in part responding to mortality statistics purportedly showing that the incidence of cancer and cardiovascular disease had increased dramatically in the twentieth century. He himself published research to demonstrate this in the case of "coronary disease."[47]

Initially, Ryle was in a minority. The British had long been highly critical of data on the rise of cancer and cardiovascular disease. In 1950 W.P.D. Logan, chief statistician in the General Register Office, provided what is probably the best and most concise explanation of why statistics about changing mortality rates could not be considered accurate, while conceding that the incidence of some diseases may have increased.[48] Several years later, Logan published the results of a long series of sickness surveys carried out from 1943 to 1952 asking monthly sample populations about recent illnesses. The report, which appeared the same year as the report of the Commission for Chronic Illness (CCI) in the United States, distinguished between illnesses that had recently appeared, that were recurrent, and that were continuing from an earlier period.[49] But neither of the latter two categories raised alarm bells or the issue of chronic illness generally.

Nonetheless, the notion of changing disease mortality and morbidity gradually spread. Whether the actual incidence of chronic diseases was rising remained uncertain in the context of generally lower mortality rates, but this was increasingly irrelevant (as it was in the United States) because their *proportion* of all mortality causes increased as infectious diseases became controllable. Also notable was the much less significant decline in mortality of one group, middle-aged men, who made up much of the subpopulation writing about disease. But even more important were changed expectations; based on recent history, people now expected medical science to do something about serious diseases, whether their incidence was rising or not. Reports about the growing burden of chronic diseases could not be dismissed as easily as in the past.

Part of this shift certainly reflected an altered international climate (discussed in chapter 8). How much was due to direct American influences, especially the work of the CCI, is less clear. The various volumes of the commission's final report were published at about the time that the World Health Organization (WHO) and some British epidemiologists were taking notice of chronic disease and were reviewed by the major British medical periodicals. A. P. Thomson, who had started the Birmingham surveys, reviewed all four volumes of the CCI's final report for the *British Medical Journal*. His accounts like those of the *Lancet* reviewer criticized the proposals for their vagueness, imprecision, and avoidance of difficult decisions. But they did not dispute that chronic illness was a critical category of healthcare and prevention. McKeown, who had grown up and received his medical education in Canada, was deeply interested in the American healthcare system about which he published a sophisticated study in 1948. (Although he said not a word about chronic illness, he cited the National Health Survey of 1935–36 and got the date wrong in the text.)[50] The influential epidemiologist Jerry Morris visited the United States in 1947 and was influenced by psychosomatic medicine, aspects of which he utilized in his early research on duodenal ulcers.[51] As an indication of how intense transatlantic cross-fertilization was becoming, a major American epidemiologist, Kerr White, credits a 1952 keynote address by Morris at a meeting in Chapel Hill for establishing a research agenda for health service research at the University of North Carolina.[52]

A critical component in the spreading view that chronic diseases had become the dominant health problem of the era was the widespread adoption of risk factor epidemiology as the basis of public health. Although risk factor epidemiology is frequently viewed as an American development, Britain had its own tradition of multifactorial epidemiology growing out of efforts to understand complex epidemics.[53] Sir Richard Doll and Bradford Hill published their first work on the link between tobacco and lung cancer in 1950, the same year as the first American publication on the subject. Jerry Morris published his study on physical activity or the lack of it as a factor in heart diseases in 1953, well before the publication of the first results of the Framingham study (although he was well aware of American research and cited it).

Morris, along with McKeown, was during the 1960s probably the leading British exponent of social medicine, the British intellectual tradition that played a leading role in transforming chronic disease into a new social problem in that country. Social medicine was established in the United Kingdom with the creation of a handful of university chairs in the field for figures like Ryle and McKeown. Morris became director of the new Medical Research Council (MRC) Social Medicine Research Unit in 1948. In 1956 the British Society for Social Medicine was founded.[54] Dorothy Porter has described social medicine as an international movement built around a rather vague notion that tied health status to social conditions and believed that the application of social science perspectives could improve both. In Porter's account, the movement focused during the interwar years on the health problems caused by poverty and inequality, as well as the creation of a national health service, but became during the postwar period increasingly institutionalized, less political, and more oriented toward pure quantitative research. Social etiology of disease remained central to the movement, but the meaning of "social" altered. Rather than referring to broad structural conditions difficult to act on, the "social" was broken down into manageable components focusing on behavior among different social groups, redefining in the process "class as a cultural rather than primarily an economic category."[55] Unhealthy behaviors were highlighted in this kind of thinking, although environmental and social factors remained prominent concerns. Now that healthcare was a social right provided by government, citizens were considered to be responsible for acting in ways that minimized the risks of disease.

The focus on health behavior became central to the "new public health" in Britain following a report in 1962 by the Royal College of Physicians on smoking that helped create a new risk-based public health in which government assumed responsibility for advising the public on health matters by utilizing the most up-to-date mass media. Rather than just providing factual information to citizens who could be assumed to act rationally, authorities sought to persuade the public of the dangers of risky behaviors. Prevention "lay within a new framework of individual responsibility . . . [and] a new framework of governmental intervention in individual behavior."[56] The sources for this new orientation were manifold: the campaign against drunk driving organized by the Ministry of Transport; American models of mass media public health advertising; and the shift from local to central authorities in health education campaigns. Although it remained largely limited to anti-smoking campaigns during the 1960s and 1970s, this strategy was eventually extended to alcohol, exercise, nutrition, and venereal disease.[57]

The agenda of the new public health, while influenced by many factors, had at its center population-based (or risk factor) epidemiology. Jerry Morris, of the Social Medicine Research Unit of the MRC, is widely credited as the major influence in the development of both this epidemiology and the new public health that emerged from it.[58] Morris's early work with Richard Titmuss looked at diseases believed to have social causes like rheumatic heart disease and duodenal ulcers. But his interest in occupational epidemiology led him to recognize the influence of exercise or the lack of it on cardiovascular disease and to the effects of individual behavior on health. His thinking accordingly shifted; illnesses once "bred in poverty and malnutrition" and the failures of the social system had been replaced by illnesses originating in high living standards and the successes of the social system. As a result, individuals now had considerable influence over their own health and needed to utilize self-reliance and self-control. "It is becoming clear that in the modification of personal behaviour, of diet, smoking, physical exercise and the rest, which look like providing at any rate part of the answer, the responsibility of the individual for his own health will be far greater than formerly. It will not be possible to impose from without (as drains were built) the new norms of behaviour *better serving the needs of middle and old age*. They will only come about in a new kind of partnership between community and individual."[59] These new behavioral norms targeted the middle-aged as well as older people and applied to a number of diseases. It was not immediately apparent, however, that the broad category, chronic illness, was useful to the aims that Morris was pursuing.

In a path-breaking paper published in 1955 that was the precursor to an even more famous book of the same title, Morris began by discussing the substantially higher rates of mortality among middle-aged men as opposed to women and the general leveling off of mortality decline among this group that he then linked to three diseases—cancer of the bronchus and coronary thrombosis, major causes of mortality, and duodenal ulcers, a major contributor to morbidity. His goal was to use these and other conditions to demonstrate the usefulness of the discipline of epidemiology. Despite these initial examples, Morris used the word *chronic* only three times in the article, in connection with mastitis once and chest disease twice. Obviously, chronicity was not a very pertinent category for his purposes.[60] Neither did the first edition of his book *Uses of Epidemiology* published in 1957 have much to say about this general category. But the very last section of his text indicates almost as afterthought that despite a different vocabulary, he was moving in the same direction as American counterparts like Lester Breslow (who also considered himself a man of the left), whose article in the *Journal of Chronic Disease* he cited in his bibliography. "One of the most urgent social needs of the day is to identify rules of healthy living that might reduce the burden of the metabolic, malignant, and 'degenerative' diseases which are so characteristic a feature of our society."[61]

By 1963, Morris was referring directly to the larger category of chronic disease. In a 1963 talk to the Royal Society of London, he discussed health problems by age group, starting with infant mortality. He got around eventually to the middle-aged, particularly men, "who since the First World War have emerged as a vulnerable group in Western Society." He gave several reasons for this, including the possibility "of a real increase of incidence of chronic disease." He then went on to discuss the "natural history" of chronic disease, a term "used today to embrace the miscellany of metabolic, malignant and mental disorders that dominate medicine in an ageing population." These diseases used up a great many medical resources and caused the great majority of deaths at all ages. "The inclusive term, 'chronic diseases,' has the advantage of emphasizing several common features: their long-drawn period of development, the often insidious onset and, frequently, the lifetime course. This last in turn has further implications in the common nature of the practical problems, physical, mental and social, that arise and the kind of care the community should be trying to provide."[62]

This, then, was a pragmatic term that allowed for wide-ranging actions that might influence the course of many different diseases. These actions could be classified into three categories that Morris would develop further in the second edition of *Uses of Epidemiology*, which more energetically embraced the term *chronic disease*. First there were those dealing with advanced cases where the goal was to prevent disability "which, it is now appreciated, is conditioned by mental and social processes, whatever the nature of the basic pathology." A second category dealt with early disease, mainly symptomless but often controllable or curable if discovered in time. Echoing American writers, Morris suggested that "this is the area of clinical medicine where automation and computers are likely to come into their own."[63] And finally there were precursors to chronic diseases, precancerous states, in particular, as well as high blood cholesterol. Whether any specific course of action in these cases was useful remained to be determined by further studies. Using the general term *chronic disease* had one further practical implication. There were, Morris thought, "*ways of healthy living*, the wisdom of body and mind and the principles of social organization that will reduce the burden of the chronic diseases and improve the quality of life."[64] One needed to avoid various dangers; some, notably air pollution, were environmental. (Later in his career, Morris researched the effects of mineralized drinking water.) Others were behavioral: obesity, lack of exercise, and smoking, as well as the more elusive psychological and social stresses that, Morris insisted, were associated with a multitude of illnesses. It was thus possible to think beyond individual diseases to a more general category, "chronic illness," for which a focus on individual behavior could become a major strategy of health promotion. By the mid-1960s, such thinking had become widespread.[65]

In another paper published in 1963 on the prevention of disease in middle age, Morris repeated much of his earlier analysis. But he also emphasized the wide variety of services that were needed by sufferers of chronic disease and he visualized MOsH acting for the community in organizing and coordinating "medical and social provision, welfare facilities, occupational facilities and so forth."[66] This need for coordination underlay what is certainly the most contentious aspect of medical reform during this period: the transformation of public health into community medicine, integrated within the National Health Service rather than local government, and turning MOsH into specialists of community medicine. Much has been written about this reform that is generally conceded to have worked poorly and demoralized public health workers for several decades.[67] But in its initial formulation by Morris, it was meant to fill several lacunae relating to chronic illness: prevention; determining what health services were necessary for each community and coordinating their activities; and ensuring continuity of long-term care, especially for individuals requiring many different service providers. In an influential article written while the reform of 1974 was being discussed, Morris emphasized the burden of chronic disease as a major reason for "the evolution of the community physician as epidemiologist, administrator of local medical services, and community counsellor, professional man and public servant."[68]

Morris was enormously influential in introducing the new understanding of chronic disease to the United Kingdom. When the *International Journal of Epidemiology* published a posthumous paper by Morris in 2010, the year following his death, the editor of the journal titled his introduction "The end of the beginning for chronic disease epidemiology,"[69] playing on Winston Churchill's famous phrase while highlighting Morris's pioneering role. But Morris was hardly an isolated figure. From the 1960s, without abandoning (or solving) the problem of elderly people, British healthcare and public health adopted, gingerly and with skepticism befitting a less affluent former empire with a publicly financed healthcare system, some of the policies that were coming to dominate American healthcare.

In 1961 the Society for Social Medicine's annual meeting included a plenary symposium devoted to "The General Practitioner and Preventive Medicine." One speaker, Maurice Blackett, called on general practitioners to become fully engaged in preventive work in order to diagnose chronic diseases in their pre-symptomatic phases while treatment was still possible.[70] In 1961–62 the steering committee of the Divisions and Study Group of the BMA chose to have local groups gather together suggestions by doctors on "Practical Steps in the Prevention of Chronic Disease." Each division or group was advised to select for study one disorder from two categories of illness; "major" (like arteriosclerotic heart disease) and "minor" (like dental disease and obesity). The steering committee also expressed the wish that "all Division and Study Groups will try to find the right answer to one general and controversial question: 'Are periodic medical examinations of the apparently healthy desirable as a means of early detection and prevention of chronic disease?' "[71] In 1963 a working group was set up in the Ministry of Health to make recommendations about "the young chronic sick"—patients ages 15–59 who did not have mental illness. The group organized a pilot survey of hospital provisions for such patients in certain regions to gauge the extent of the problem. It was determined that much more information was needed in order to make proper support services available.[72]

The Limits of Chronic Disease Public Health in the United Kingdom

Despite this visible interest in chronic disease, some caveats are in order. First, it is remarkable how few of the major reports produced by the Ministry of Health or the Department of Health and Social Security treated chronic disease as a distinct problem. Discussion tended either to be very general or narrowly disease- or population (elderly)-specific. Part of the problem was operational. In 1968 the Ministry of Health reintroduced prescription charges (or user fees) for medications. (This charge had been eliminated several years earlier.) But it exempted four categories of patients: children under 16 years of age; elderly patients over 65; expectant mothers; and the "chronic sick." There was an immediate backlash from doctors, who called the plan "stupid," "shocking," and a "completely useless scheme." Specifically, practitioners took issue with the provision for the "chronic sick," arguing that it was impossible to effectively define such a category and that placing doctors in the position to determine what patients qualified for the exemption would "damage the doctor-patient relationship beyond the point of no return." The General Medical Service Committee of the BMA took the stance that it was up to the government and not the doctor to decide who should or should not pay the prescription tax and repeated its earlier position that "chronic disease was not definable."[73] The minister of health, Kenneth Robinson, recognized the problem and in a memorandum to the cabinet admitted: "The chronic sick present great difficulties because it is not possible to produce a reasonable definition of who should be included in the exemption or to identify them . . . To prepare a list of ailments as a definition of chronic sickness would be unfair to others whose illness, though intermittent, might lead to hardship. To try to identify them by reference to length of time off work would be unfair to, e.g. the housewife."[74]

Ultimately, it was decided to replace the category of chronic sick with a list—remarkable for its brevity—of specified medical conditions requiring continuous medication. This included various permanent fistulas, diabetes and other endocrine disorders requiring substitution therapy, epilepsy requiring continuous anticonvulsive therapy, and continuing physical disability that made it impossible to leave one's house unaided.[75] By October 1969, it was calculated that forty-four thousand exemption certificates had been issued to patients suffering from such specified medical conditions.[76] The list of exempt conditions underwent considerably scrutiny over the years, and was often criticized as "inequitable and anomalous." In 2009, while other parts of the United Kingdom abolished the prescription tax entirely, the NHS in England added cancer to the list of exempted conditions.[77] What emerges most clearly in this story is that while chronic illness was a relevant concept for certain purposes in the United Kingdom, defining it for concrete program benefits was almost impossible.

This, however, is true for any system of reimbursement that tries to give special rights to hard-to-define categories of patients.[78] The absence of attention to chronic disease in the United Kingdom had deeper roots. The essence of the new public health was transforming behaviors in large measure to prevent chronic illnesses. The problem is that no one really knew how to effectively promote such activities, especially in the absence of funds and local institutional structures. By the beginning of the 1980s, the failure of health promotion was beginning to be clear. An article of 1981 pointed out that the failure of prevention had been discussed in numerous government reports and white papers. "However, none of these Government reports considered the main problem in prevention—the lack of an effective public health service in Britain to put it into practice."[79] This absence did not just have a negative impact on prevention policies. When one takes account of just how much effort the US Public Health Service put into dealing with chronic illness, it becomes clear that without a powerful public health presence, the United Kingdom was missing a key component of the American chronic disease coalition.

Success or not, focus on modification of individual behavior fit less comfortably into public health thinking in Britain than it did in the United States. The British public health tradition included a longstanding concern with health inequalities based on social class. Modifying behavior was undoubtedly easier to sell to governments than social change, especially when promoted by a confirmed man of the left like Morris, but the social tradition continued to be influential. In 1980 the famous Black report reconfirmed its central importance.[80] This tradition had little effect on public policy until the 1990s, but it did have an effect on health research, in combination with the famous Whitehall studies.[81] It also had an effect on public rhetoric. In a 1981 article, Morris still talked about chronic diseases but he devoted considerable space to health inequalities and was frank in his admission that aside from smoking, little was known about the effects of behavior change or disease causes.[82] By the end of the 1980s, moreover, prevention of communicable diseases was again on the public health agenda.[83] Whatever else one can say about it, discussion of preventive health policy in Britain was hardly monolithic. When a report appeared in 2004 emphasizing the importance of individual behavior and choice, the response of the UK Public Health Association was scathing. "We question whether this is in fact a *public health* White Paper at all in the widest sense of the term. It smacks of an old fashioned, medically dominated, health education approach that fell into disrepute some years ago because it has been seen to be completely ineffective."[84] Within this social tradition of public health, chronic illness was real enough but hardly a central issue.

But less central does not mean absent. The British turned out to be selective about the aspects of American chronic disease policy that they were willing to adopt. Periodic physical examinations—except for schoolchildren—did not become widely accepted in the United Kingdom or in much of the rest of Europe. At a meeting of the BMA in 1965, there was what appeared to be a not tongue-in-cheek debate on the amendment: "In the opinion of this house, the routine medical check-up in the over-forties does more harm than good." There were speakers on both sides of the question and in the end the chair asked that no vote be taken "since if the motion were passed the impression might be given that medical examinations were harmful. On the other hand, if it failed, doctors might be represented as wanting to carry out routine check-ups for their own personal and financial advantage."[85] A somewhat positive article on the subject in 1981 provoked a storm of critical letters to the editor.[86]

Multiphasic screening, widely recognized as an American import, received considerable attention.[87] This was not an entirely new subject in the United Kingdom. During the 1950s and early 1960s, there was much debate about the usefulness of programs of mass screening for tuberculosis.[88] In 1962 J.M.G. Wilson, senior medical officer in the Ministry of Health, undertook a WHO-sponsored trip to the United States to study screening and early detection of disease. He visited fifteen cities and spoke with ninety-eight experts throughout the country. His conclusions were mixed and, unsurprisingly, he did not think that the structure of medicine in Britain allowed for the kind of experimentation with its "false starts" that was occurring in the affluent United States. Above all, he thought that since healthcare was free, general practitioners were in the best position to use simple screening techniques on their patients. His final conclusion was that nothing could be determined without "much thought, effort and experiment."[89] Wilson would go on a few years later to co-author the WHO guidelines on screening and became the major expert on the subject in the United Kingdom.

Experimentation on a small scale did, in fact, take place. In 1962, in the county borough of Rotherham, the MOH started a multiple screening service that proved extremely popular. By 1965, the clinic had eleven test stations for anemia, chest x-ray, diabetes, vision, glaucoma, hearing, arteries and heart, lung function, cervical cytology, breast cancer, and mental health. The MOH admitted that there were many problems with such tests and no proof that they affected morbidity or mortality rates. Nonetheless, he concluded: "I think that the onus of proof rests with the other side. So far as we are concerned in public health, I do not think we can afford to wait and see because to stand still in a changing world is to retreat."[90]

In July 1965 Jerry Morris chaired a colloquium on "Surveillance and Early Diagnosis in General Practice." The two chief questions were screening programs and the possibility of detecting diseases in general practice. The first remarks were by J.M.G. Wilson, who reiterated the ten rather stringent conditions he had developed for the WHO that needed to be met in order to justify screening programs. (These included accurate tests, effective therapies in case of positive results, and existence of a recognizable latent or early stage during which intervention made a difference.) M. F. Collen was brought in from California to talk about the automated multiphasic tests he had developed. Professor G. Jungner talked about an automated blood-screening program carried out in Sweden. Despite many questions, there was general consensus that experiments with screening should continue and be expanded.[91]

Not everyone agreed. Part of the problem was that in the absence of public health or administrative structures for screening, the task fell to general practitioners, who frequently felt they had better things to do. "The time of doctors is too valuable to waste on examining the transparently hale and hearty when there is an unfailing supply of those who are really sick."[92] Perhaps the most common view was expressed by one of the pioneers of what would become evidence-based medicine, Archie Cochrane, who expressed great faith in the future of screening if proper clinical testing was done but who remained unwilling to invest heavily in mostly unproven and expensive programs.[93] Indeed, the Department of Health and Social Security supported financially "a not inconsiderable" volume of research on the assumption that it would become increasingly popular and that basic questions about its validity and value needed to be addressed. Results, however, proved to be disappointing. The Social Science Unit of the Rotherham Health Department evaluated its very popular multiple screening program and concluded that the costs were "prohibitively high for the relatively few ascertained benefits it conferred."[94] Of the two major studies that seriously questioned the efficacy of multiphasic screening, one was conducted in Britain.[95]

In the end, the British like the Americans abandoned multiphasic screening in favor of a few disease-specific programs. There were numerous local experimental screening programs, particularly for breast and cervical cancer. The NHS ended up adopting a few large-scale screening programs that, while hardly uncontroversial, enjoyed significant public support and could at least be defended with some evidence of efficacy. By 1970, the NHS provided comprehensive screening for cancer of the cervix and phenylketonuria, although these programs were not necessarily effectively organized. In 1986 a committee convened by the minister of health presented a report supporting mammography for all women over 50; two years later, the NHS Breast Screening Program was established and began screening women every three years. That same year the national cervical screening program, in existence more than twenty years but recognized to be ineffective, was re-launched.[96]

Unlike the problem of older people that remained at the forefront of policy discussions, the more general notion of "chronic disease" that emerged in the United Kingdom remained for the most part a background issue, a convenient shorthand for talking about a range of different illnesses and comorbidities. It was part of the thinking behind several policy changes that took account of the growing significance of certain diseases. The new public health, however, did not identify itself primarily as a response to chronic illness. Community medicine never tried seriously to coordinate care of the chronically ill. Articles and meetings that discussed health education often cited American data and took for granted that diseases being targeted were different from those of the past but seldom limited their concerns exclusively to chronic illnesses.[97] A National Health Service that dealt with many different kinds of patients and was relatively underfunded in comparison to healthcare systems in other major nations could not devote great resources to one specific category of diseases. I have found no references to administrative units devoted exclusively to chronic illness, comparable to the many departments and units that proliferated in the United States. Academic and research units did exist. But these, it was charged in 1981, had no service functions within the NHS and little influence on either doctors or local authorities.[98] Seven years later, not much had changed, according to a committee studying the public health system. "We are conscious that there is no body in the field of noncommunicable disease equivalent to the PHLS (Public Health Laboratory Service) and CDSC (Communicable Disease Surveillance Centre) with responsibility for long-term surveillance of conditions such as cancer, stroke, and cardiovascular disease."[99]

Some of this may have had to do with widely recognized structural weaknesses of British public health as it morphed into community medicine accompanied by considerable functional confusion.[100] A lot clearly had to do with the fact that a nation with a limited health budget remained preoccupied with providing adequate acute hospital care while keeping costs manageable.[101] One casualty of this focus may have been rehabilitation medicine. As in the United States, the beginnings of this field long preceded recognition of a chronic disease "problem" and were connected with the needs of the military. Rehabilitation came into its own during and after World War II when the country had to deal with both military and civilian casualties and industrial accidents. Between 1943 and 1946, the number of hospitals with specialized rehabilitation units increased from 35 to 121, and those offering specific services (like orthopedics or physiotherapy) rose even more sharply.[102] For several decades after the war, rehabilitation dealt mainly with conventional physical handicaps and was associated with vocational and counseling services in order to get people back to work. During the 1970s, somewhat later than in the United States, it expanded beyond these areas to such conditions as arthritis, respiratory disease, heart disease, and stroke. Geriatrics also maintained a consistent focus on rehabilitation with the aim of keeping elderly people as autonomous and self-sufficient as possible. Both were included within the National Health Service and were thus free to all. In 1983 the British Association of Rheumatology and Rehabilitation split into two separate entities, but it was not until 1991 that rehabilitation became a distinct specialty recognized by the Ministry of Health. By then, numerous commissions had examined the nation's rehabilitation services and a consensus had emerged that the situation had deteriorated relative to other European countries.[103]

Rehabilitation in the United Kingdom never served as a central element in a rhetorical campaign to deal with "chronic illness" as it did in the United States. In fact, references to chronic illness are few and superficial in the vast British literature on rehabilitation. The social problem that the specialty addressed most directly was disability, which might or might not be the product of disease. It is rare to read anything about rehabilitation that did not identify it with disability.[104] There is nothing irrational about this. Many American practitioners of rehabilitation medicine undoubtedly also saw their task primarily as ameliorating disability. But the relative lack of British works comparable to those of Howard Rusk linking rehabilitation to chronic illness, or an institutional equivalent to the American Public Health Association's Committee on Chronic Disease and Rehabilitation[105] is nonetheless striking. Rehabilitation in the United Kingdom was not about disease processes but about their consequence, disability.

The British framed many of the issues that troubled Americans in terms of disability. Characteristically, surveys looked mainly for functional disability with specific illnesses sometimes noted as a cause. A questionnaire survey in North Lambeth begun in the late 1960s included chronic disease in the title but counted disability. The reason given was "the conceptual and practical difficulties" involved in determining the prevalence of chronic disease. The study found a prevalence rate of disability among those aged 35 to 74 of 7.2% for men and 9.7% for women. Chronic respiratory disease, mainly bronchitis, was the single most common condition associated with disability. But no consequences were drawn from this finding; the goal was clearly to count something that was quantifiable and the authors went on in other publications to develop measures of disability.[106] The most important government-sponsored survey of the period, carried out in 1969 and published in 1971 and popularly called the Harris Survey, focused on "disability and impairment." It included detailed data showing that the vast majority of such disabilities were the result of chronic diseases but again little was made of this finding.[107] It was the consequences rather than causes of disability, in this case the need to determine claimants for disability benefits, which were subject to examination.

In addition to considerations of methodological rigor, politics deeply influenced the emphasis on disability. In 1970 a newly elected Labour MP named Alf Morris introduced a private member's bill in Parliament, the Chronically Sick and Disabled Persons Act, which focused primarily on disabled individuals, as did the media attention around it. The act imposed certain responsibilities on local welfare authorities to improve support for the disabled. Authorities were enjoined to determine through surveys the number of disabled in the community; provide or adapt housing for them; and offer services like meals at home or community centers, aid to travel, adequate access to public buildings, toilets, and the like. One provision of the bill was that young patients were to be separated from elderly patients in hospital wards and public residential accommodation, a provision that had particular relevance for mental health disorders.

Alf Morris went on to become the first minister for the disabled. In 1973, following public discussion of compensation for thalidomide children, a group of sixty-eight eminent individuals wrote an open letter to Prime Minister Edward Heath demanding improved pensions and services for the disabled. In a long commentary to the open letter, Peter Townsend and Walter Jaehnig discussed the complex issues involved in implementing the act of 1970, particularly the government's weak commitment to it, but ended with a final paragraph arguing for a more comprehensive policy that among other things obliquely suggested that greater attention be paid to the disease causes of disability. "Finally this policy should address itself not only to physically handicapped people but also to the more intractable problems implied by chronic sickness, mental handicap and psychiatric disorders, conditions often left out of discussion of disablement."[108]

Researchers specializing in specific diseases had their own reasons for questioning disability policy. In 1978 a group associated with the Arthritis and Rheumatism Council Epidemiology Research Unit at the University of Manchester reanalyzed the data in the Harris Survey in terms of disease causation. They found that arthritis and rheumatoid arthritis were the most important cause (29%) of severe and very severe impairment in the "young" (under age 60). That these "forgotton [sic] disabled" needed more services was the obvious conclusion."[109] But one of the authors went somewhat further when speaking at a symposium devoted to chronic illnesses of the young in 1978. He again emphasized the disease causes of disability, particularly arthritis, but this time concluded in a way that echoed the arguments of American chronic disease reformers. "Studies are now needed of the natural history of particular impairments and disabilities—as a means of identifying ways in which we can intervene in the sequence, preventing the development of disability when impairments are present, and reducing the disadvantage that may ensue."[110]

Despite such occasional comments, the gap between disability and chronic illness grew wider with time as the increasingly militant disability movement in Britain as elsewhere rejected medical definitions of disability in favor of social and political definitions that focused not on medical causes or even resulting handicaps, but on the hardships caused by society's failure to accommodate these conditions adequately.[111] Much the same movement developed in the United States, but there disability advocates were distancing themselves from a by-now well-entrenched chronic disease movement that retained its identity and influence. In the United Kingdom, disability activists faced no comparable movement to distance themselves from.

Conclusion

From 1960 on, the meaning of *chronic disease* expanded in certain circles and took on some of the broad connotations and public health implications that defined the term in the United States. The British adjusted their healthcare institutions and research activities to take account of new disease patterns and preventive strategies. But the concept "chronic disease" did not assume a central role in health policy. It was part of the background assumptions on which programs and activities were based rather than a central object of concern. A no-longer rich nation supporting a tax-based National Health Service providing free physician and hospital care to a heterogeneous population had more immediate issues to face than an open-ended problem like chronic illness. Public health focused on specific diseases and how to prevent them, particularly through media campaigns that sought to modify behavior. The term *chronic disease* was for much of the twentieth century used primarily in connection with care for elderly people, a problem in the United Kingdom that has resisted solutions proposed by numerous commissions, white papers, and legislative reforms.

CHAPTER TEN

Maladies chroniques in France

When I first began researching chronic disease in France, I encountered two difficulties. First, French historians and sociologists had no idea what I was talking about. It was not that they did not understand the literal meaning of the words; these simply had little relevance to healthcare, as they understood it. Nor was the term used very much. Ngram graphs that show a significant rise throughout the twentieth century in the use of terms *chronic illness* and *chronic disease* in the large numbers of English-language books that Google has digitalized show in contrast a flat line for *maladies chroniques* or any related terms in French works during the century. A second and related problem is that there is an almost complete lack of historical writing on this subject. With the exception of a small number of French and North American works that I cite in this chapter, I was forced to rely on research on more-or-less related subjects. This situation is particularly astonishing given the many books and articles on chronic disease published in France during the nineteenth century. This chapter analyzes the institutional logic that made the term increasingly irrelevant for healthcare policy in the twentieth century.

Indigence and Chronicity

As was the case in Britain, chronicity in nineteenth-century France was associated with indigence, particularly and increasingly among elderly people, and was most frequently used as shorthand to designate those who should *not* be in hospitals because medicine could do little for their conditions. Three conditions in particular fell into this category: old age, infirmity, and incurability. These rendered the very poor eligible for public assistance or charity since they made work and economic self-sufficiency impossible and were not associated with moral failings. (Foundlings were another important group whose blameless helplessness made them eligible for assistance, but they are not relevant to this discussion.) While these conditions permitted "assistance"—food, lodging, small subsidies, and the like—they frequently excluded sufferers from the medical care that the indigent in many localities had a right to as consequence of a law passed in 1851. Since space was limited in hospitals that provided medical care, it was thought proper and appropriate to reserve them for those who would respond to treatment: the sick, or *les malades*, in the parlance of the period. Those not considered *malades* could, however, be lodged in nonmedical institutions, hospices, which generally fell under the same municipal administration as hospitals (in Paris what became the Assistance Publique ([AP] in 1849 and in Lyon the Hospice de Lyon). Only a few localities could support two separate kinds of institutions and the majority had mixed institutions called hospital-hospices.

Although categories were not always clear-cut, official statistics in 1851 indicated that in all of France there were 337 hospitals, 199 hospices, and 734 hospital-hospices. This translated into 63,237 hospital beds and 55,052 hospice beds. In Paris, general hospitals contained 3,783 beds; specialist hospitals (for sick children, venereal diseases, and the like), 3,429; and hospices and retirement homes, 10,691. There were several huge hospices in Paris that had mixed vocations. Bicêtre and Salpêtrière, each with more than 3,000 beds, were predominantly hospices for elderly men and women respectively. Each also had a section for mental illness, comprising about one-quarter of all beds. In both institutions there were also small numbers of younger incurably ill patients, primarily sufferers from tuberculosis, cancer, and what we would now call disabilities. Two other large Parisian hospices, one for men and one for women, were devoted to "incurables" and together held a little more than 1,200 beds. Despite the large number of beds in hospices, the lengthy stays of many patients meant that annually about 12,000 elderly, infirm, and incurable patients were housed in Paris while about 99,000 *malades* were in hospitals. Nationally, at midcentury, there were about 650,000 *malades* annually in hospitals and about 100,000 individuals in hospices.[1] The latter moreover could be of several different sorts. Bordeaux, for instance, had two hospices, one for elderly and another for incurable patients.[2]

These complex categories give some idea of how unstandardized the system was, a feature intensified by local variations. Although reserved officially for the indigent, both hospitals and hospices contained small numbers of paying patients throughout the nineteenth century. Even the basic distinction—*malades* in hospitals and "elderly, incurable, and infirm" in hospices—was not always respected because hospital physicians would not always turn away individuals who had nowhere else to go. It was common, for instance, for men and women to wait for years before getting a place in one of the two hospices of Bordeaux.[3] Doctors frequently admitted tubercular patients a few days before death in order to spare families burial expenses without tying up beds for long periods. An intermediate category that would eventually be termed "chronic" (in the 1893 law on free medical care for indigents) were not incurable or likely to die soon and thus had a right to hospital care even if there was little medicine could do for them. These "chronics" were frequently perceived as inundating hospitals and taking up beds.[4]

The pressure to develop medicalized hospitals focusing on treatment was uneven but intense. Medical care, it seemed, could actually cure some people and get them back to work. A ministerial report to the king in 1837 lamented the large sums spent on hospices since it allowed the poor to think that relief in old age was guaranteed and thus thwarted the spirit of work, thrift, and saving. The report suggested that some hospices should be eliminated so that more could be spent to keep the elderly and sick poor at home with their families. To this classical liberal view of welfare assistance was added a more innovative suggestion: that part of the revenue going to hospices should instead go to hospitals since these more useful institutions were currently unable to admit all the sick that might benefit from medical care.[5] But this was an unusual view since part of the problem was that incurable and chronic patients were in fact taking up space in urban hospitals; getting them out by setting up new hospice-like institutions in the suburbs or countryside was a regular theme of discussion.

Advancing medicalization of hospitals made Paris the leading medical city of Europe during the first half of the nineteenth century, whereas hospitals in Lyon, France's second largest city, were slower to abandon traditional care of indigents.[6] Medicalization did not necessarily mean extensive medical procedures. Most hospital funds were spent on food and lodgings, considered part of treatment for curable conditions. The cost of medical and nursing services rose only slowly during the century.[7] One result of this pressure to medicalize hospitals is that admissions to hospices nationally seem to have stagnated during the century and actually declined during the 1870s and 1880s to around sixty thousand to sixty-five thousand annually.[8] Aside from elderly people, those who suffered most from exclusion from medicalized hospitals were "incurables," particularly those diagnosed with tuberculosis, by far the most common disease of the nineteenth century. Those suffering from advanced cancer were less numerous but deeply disturbing to contemporaries because of frequently unpleasant symptoms. The Paris hospital system's central admitting service that assigned patients to appropriate institutions devoted considerable effort to keeping "incurables" out of hospitals. Rules might be broken and the latter admitted, but with the likelihood that they would be dismissed whenever beds were needed. Local welfare bureaus with their meager resources and wealthier Catholic philanthropies might aid incurable or elderly individuals in cash, kind, or lodging. A Catholic order, Les Petites Soeurs des Pauvres, providing hospice care to the elderly poor, doubled in size during the Second Empire. In Lyon another religious order, Les Dames du Calvaires, founded in 1842, staffed hospices and spread from Lyon to five other French cities.[9] By the Third Republic, the Catholic charitable sector was diverse and expanding while the public system, also considerably dependent on private donations, was, in Colette Bec's phrase, like "a poor relation."[10]

During the early Third Republic, the government carried out a survey of welfare provisions, including hospital and hospice facilities, which showed serious insufficiencies. In Paris the central admission service was widely criticized for its cavalier treatment of chronically ill individuals who were regularly sent away and told to return later. Others criticized Parisian hospitals for accepting too many incurable and infirm patients and worsening the problem of overcrowding. The solution was to build more hospices. In 1869 the residents of the men's and women's hospitals for "incurables" were moved to a two thousand–bed hospice in the suburb of Ivry. In 1876 the new Laennec Hospital officially replaced the two incurable hospitals in central Paris. It was largely devoted to tubercular patients but gradually evolved into a general hospital with a large tuberculosis ward. That same year Parisian authorities began offering tiny pensions to terminally ill "consumptives" who lived at home, as they were already doing for cancer patients and epileptics.[11]

Following the economic depression lasting from the 1870s to the 1890s, and with the specter of socialism and trade unionism looming, the need to reform assistance became more pressing. During the 1890s, successive French governments introduced a series of laws that were to become the basis of the welfare state, justified by the doctrine of solidarism, which preached state action to promote social solidarity. A law passed in 1893 mandated that free medical assistance be provided to all indigent persons. This assistance included hospital treatments and consequently the number of hospital patients increased significantly during the following decades. After some debate, it was decided that certain categories of chronic disease (officially recognized as *maladies chroniques*), including early-stage consumption, could be treated in hospitals. But the moment a patient was judged incurable, the government stopped paying hospital costs.[12] A series of laws, the

Charter of Mutual Help Societies (1898), the law on industrial accidents (1898), and the law introducing pensions for industrial and agricultural workers (1910), along with social protection plans developed in certain large industries, meant that more and more working-class men could benefit from paid medical and hospital care.[13]

There were those who thought that the "elderly, infirm, and incurable" deserved similar rights. Anticlericalism came into play as well since hospice care had a strong association with Catholic religious orders and charities. In 1905 a law was passed providing obligatory assistance to the indigent "elderly, infirm, and incurable ". Starting in 1907, communes had to provide a small monthly stipend to all those meeting conditions of indigence and residence. Hospices had to take such people in for free if they were incapable of living alone. The national government paid more than half the costs of the program with local authorities paying the rest. The funds were used to create publicly funded nondenominational hospices. The legislation also included appropriation of revenue from existing hospice programs.[14] The effects of these laws were uneven but significant. Between 1869 and 1908, the number of hospital beds increased by 32% while the number of hospice beds rose by 41%. By 1911, nearly 580,000 people were assisted either at home or in an institution.[15]

The Twentieth-Century Burden of Chronic Disease in French Hospitals

During the interwar years, health reform continued in the context of what appeared to be a crisis of population aging. This was attributed to the combination of low birth rates, military and civilian casualties during World War I, and mortality rates seemingly higher than those of France's European neighbors.[16] The nation's hospital system was expanded and modernized and gradually became the center of medical care.[17] The number of those with a right to hospital care increased again in 1919 due a law granting war pensions. The formal limitation of hospitals to the poor continued to be debated with no immediate resolution. As a result, a network of private hospitals and clinics for paying patients began to spread.[18] But even if universal access to hospitals was not legislated until 1941, the reality is that French hospitals were in large measure accepting paying patients during the interwar period. The introduction in 1930 of a system of social insurance covering the risks of illness, disability, maternity, and old age for employees in commerce and industry added to the growing availability of third-party payers.[19] The number of individuals admitted to hospitals and hospices thus rose from around 400,000 in 1883, to 741,000 in 1913, and to more than a million in 1938.[20]

One final development influenced the expansion of hospitals: the shrinking notion of incurability. The division between hospitals for *malades* and hospices for the "elderly, infirm, and incurable" remained a major organizational principle.[21] Nonetheless, the groups cared for in the latter institutions changed as the category of incurability gradually contracted. Calls to admit epileptics to hospitals became persistent during the last decades of the nineteenth century and tuberculosis was increasingly perceived as treatable; in the early twentieth century hospitals admitted a large number of tuberculosis patients and a national system of sanatoria was (belatedly) set up. Cancer became a treatable disease with radiation during the 1920s; a national system of cancer centers was created and better-equipped hospitals began to treat it as well.[22] Diabetes became a controllable disease from the 1920s on. As the notion of incurability contracted, elderly infirm people became increasingly the major clients of hospices. The number of individuals in such institutions (less than one hundred thousand before World War I) rose during the interwar decades.[23] Numerous hospices were constructed or renovated. One historian has called this the "golden age" of hospices for older people, in spite of strict institutional rules limiting residents' freedom and minimal medical care.[24] From 1929 to 1939 the number of beds for elderly patients rose far more quickly than did the number of beds reserved for *malades* (56% versus 12%).

Despite renovation and new construction, a growing urban population, expanded patient access, and lack of medical care in hospices meant that chronic patients continued drawing criticism for taking up acute-care hospital beds. By 1928, administrators of the AP were complaining that Parisian hospitals were seriously overcrowded and held three thousand beds beyond what buildings could safely accommodate.[25] It was decided to build at considerable expense four large general hospitals in the suburbs of Paris (now included within the catchment area of the AP) to house four thousand patients. The original plan was modified because local authorities wanted smaller, more community-oriented institutions, and one of the hospitals eventually built was devoted specifically to chronic care; it was located in the then remote suburb of Garches west of the city. The goal was to get patients suffering from either lengthy or incurable diseases out of acute-care beds in the central Paris hospitals. Because of disagreements with the communal authorities, construction of the Hôpital Raymond Poincaré was not fully completed until 1936, when it opened with 1,200 beds. Unlike the Hospice de la

Reconnaissance, a traditional hospice that was literally next door (and that would eventually be annexed to the hospital), Poincaré had limited medical facilities. It was divided into three sections, each directed by one physician; the assumption was that patients were stable and did not need frequent physician contact. There was only a room for light surgery (there was no operating theater), and there were no outpatient clinics. Physicians had to fight the administration to obtain minimal laboratory facilities. As a result, the hospital was cheap to run, about half the cost of a general hospital. Efforts were made to create a pleasant environment within these almost rural surroundings but some of the common rooms had as many as sixteen beds, a situation that was no longer considered acceptable. (Four beds to a room was the new norm.) In the late 1940s and early 1950s, the institution changed course due to pressure from the Ministry of Health. First, a rehabilitation program was introduced for childhood polio victims and eventually expanded to adults while a job-training program for patients who had joint and skeletal problems was established. The hospital eventually became a full-scale rehabilitation center.[26]

The creation of the modern social security system in 1945 further expanded the clientele of hospitals and intensified pressure to rationalize the system. The insurance funds in particular demanded stable categories of patients with fixed daily rates on which they could base reimbursement. Chronically ill patients were especially perplexing because no one knew exactly how to define them, what care they had a right to, and with what kind of insurance reimbursement. In 1949 the director of hospitals in the Ministry put these questions to some of his officials, who seemed equally confused about the fact that 2,306 Parisian hospital beds were reserved for chronic care. It turned out there was little legal basis for the existence of these services and little agreement about what illnesses they covered; it seemed as if the only point of unanimity was that tuberculosis was not considered a chronic disease. "For the rest, imprecision, uncertainty."[27] The closest to formal administrative status was provided by a ministerial decree of 1943 that fixed the daily cost (*prix de journée*) of chronic and convalescent beds. A very partial survey of chronic patients in the AP (one hundred cases) concluded that 92% of those who should have been in hospices rather than hospitals were "elderly, infirm or incurable." One administrator estimated that old people (*vieillards*) filled 24% of all hospital beds in France.[28]

In 1950 the National Council of Hospitals in the Ministry received a half dozen reports from administrators and physicians on services for chronically ill or older patients and then a synthetic report was produced.[29] This report defined chronic disease as "a durable pathological state" whose evolution could be slowed, stabilized, or improved by effective therapy. The approach was nothing if not traditional. For practical reasons, mental illness and tuberculosis were not discussed. Chronic patients were divided into three categories. The non-recoverable (now termed *non-récupérable*) required minimal medical therapy. The ameliorable included sufferers of neurological and cardio-renal problems and required considerably more medical resources. The recoverable (*récupérable*), included victims of accidents, polio, and cardiac insufficiency, usually under 60 years of age, whose cure was "foreseeable and certain" with adequate therapy, thus demanding considerable medical resources and personnel. In addition, it was proposed that geriatric services could be established by collecting together existing but scattered beds for old people (*vieillards*).[30]

In the end the administration of the AP followed a two-pronged policy. First, existing institutions like the one in Garches took on new functions, primarily rehabilitation. Second, the policy of moving chronically ill patients to the suburbs intensified. In 1954 the administration of the AP set up another chronic-care hospital in the southeastern suburb of Creteil. In a hurry to get started, it utilized a new technique, construction with prefabricated parts. By the time all the buildings were completed in 1959, the Hôpital Albert-Chenevier held about seven hundred beds devoted in most cases to elderly patients, who required a lengthy period of care but not urgent or highly specialized therapy. Following the newest criteria, no rooms in this institution held more than four beds.[31] In 1962 the hospital opened up a rehabilitation unit and in 1969 a psychiatric ward.

A set of reforms in 1958 and 1959 opened the modern era of medical education in France by creating university hospital centers, and also amplified the role of hospitals to include diagnostic and preventive examinations, as well as functional rehabilitation. The reforms opened up a private-patient sector for hospital physicians and surgeons and created five categories of hospital institutions, including *retirement homes*—a term that replaced *hospices*, a word with very negative associations. The new term does not seem to have either stuck or raised the status of these institutions. There was no mention of chronic-care hospitals in this classification scheme. Gradually during the 1950s, Bicêtre and Salpetrière, traditionally hospitals for psychiatric, elderly, and incurable patients, evolved into general hospitals in order to deal with growing numbers of acute-care patients and to serve as teaching hospitals.

In 1960 several new chronic-care services had been set up or expanded in southern suburbs of Paris—notably the tuberculosis Sanatorium Joffre transformed into a chronic-care hospital. But this was not nearly enough to deal with the growing population of greater Paris. The administration of the AP estimated that fifteen thousand new beds were needed. In 1962 desperate officials in the Ministry of Health offered to help subsidize the creation of two thousand chronic-care beds in suburban Paris with the goal of freeing beds in acute-care institutions. The Ministry offered a grant of 5 million francs and a loan of 45 million francs to be borrowed from the state-controlled bank, the Caisse des Dépôts. The AP provided the land. Beds were to cost about half of what they cost traditionally, thanks to "modern" construction methods. But these hospitals were not to be simple dumping grounds. Both nursing care and rehabilitation services were to be amply provided. Housing for staff was to be built using yet another bank loan. In this way, three new suburban chronic-care hospitals were constructed during the 1960s: the Hôpital Dupuytren in Draveil, south of the capital; the Hôpital Charles Richet in Villiers-le-Bel to the north; and Hôpital René Muret in Sevran to the northeast. In the planning discussions, specific diseases were not at issue. Architectural questions hinged on function and particularly on the proportion of patients who were elderly; completely handicapped; bedridden; senile; or had special feeding needs.[32] In order to further relieve congestion, the Parisian AP in 1957 introduced a system of homecare. Although it was inspired by the Montefiore Hospital homecare system, private practice physicians rather than hospital physicians supervised patients at home following hospitalization. A year later, another homecare unit was created at the Gustave-Roussy Cancer Institute. In 1961 an agreement was signed between a major health insurance fund and the AP; by 1976, fifteen homecare units were in existence.[33] But the system remained small and relatively undeveloped.

Paris was probably the only French city during this period that received state aid to systematically move large numbers of chronic patients out of the city center and into the suburbs. Lyon invested in several high-tech specialty hospitals and opened a geriatric center in 1960 and a rehabilitation hospital in 1969.[34] One widely available option was to recycle tuberculosis sanatoria, no longer thought necessary, into chronic-care hospitals or hospices. Well into the 1960s, this development deeply worried tuberculosis specialists, who feared that the disease threat remained serious enough to warrant the continued existence of sanatoria.[35] Everywhere, hospices and retirement homes were built, renovated, and/or modernized by local authorities or private interests. A particularity of the French system is that such institutions were frequently managed by hospital administrations, a heritage of their traditionally mixed medical and welfare responsibilities.

In 1962, 9,968 hospices or retirement homes were run by local hospital administrations; there were 865 chronic services or wards located within individual hospitals and 1,320 hospice institutions not hospital-related in any way.[36] The number of beds in hospices and retirement homes rose from 247,000 in 1960 to nearly 340,00 in 1972. More than one-quarter of these beds were in the private sector. In the public sector, the number of hospice and retirement home beds rose by 40%.[37] Nonetheless, the problem of elderly patients in acute beds did not abate. Estimates of the proportion of hospital beds they occupied ranged from 15 to 20%.

Chronic patients were thus a very real category for hospital administrators, referring to those—mainly but not exclusively elderly—who needed to be removed from acute-care institutions and who might need special and usually cheaper forms of care. Rather than being a major social issue as they were in the United States, these patients were an irritating administrative inconvenience, one that could be resolved by administrative decisions to park them in less expensive suburban institutions: chronic-care hospitals if they needed some medical care and hospices if they did not. In 1974 the AP calculated the *prix de journée* (the daily cost for each patient's stay) of chronic patients at 139 francs, a little more than half the cost of a patient in an acute-care general ward and one-third that of a surgical patient. (Patients in specialized wards had an even higher *prix de journée*.) The figure for chronic patients was about 65% higher than the cost of elderly invalids in hospices or retirement homes (84 francs). The differences in cost reflected differences in quality of care, particularly the ratio of doctors, nurses, and other care personnel per patient.[38] Perhaps because they were so badly rewarded for their essentially custodial work, chronic-care hospitals were eager to specialize in such areas as rehabilitation and geriatrics (often both) as these fields gradually became recognized and better reimbursed.

Dissident Voices

Not everyone viewed the problem of chronicity as one of hospital overcrowding. In 1963 the French Society of Psychosomatic Medicine held a conference on the psychosomatic aspects of chronic disease. This was the certainly the event in France that came closest to manifesting American-style interest in understanding the psychological mechanisms and effects of chronic disease.[39] There was, however, no significant follow-up to this meeting. At roughly this time, the policy of segregating "chronics" in suburban institutions with minimal medical facilities began to provoke opposition. There was hardly a groundswell of protest but a handful of medical figures publicly opposed the policy. A series of editorials on the problems of Paris hospitals published in 1963 by M. Grivaux in a prestigious weekly medical journal had much to say about the shortcomings of government hospital policies but initially said little about chronic care. The announcement that new suburban chronic-care hospitals would be constructed finally provoked a reaction. Grivaux's fifth editorial ended with a brief but stinging critique of recent chronic-care policy. It cited a speech by Dr. J.-A Huet, a founder of gerontology in France, and recent articles by Dr. Henri Péquinot criticizing the plan. All doctors, the editorial went on to say, must associate themselves with these views. It was unacceptable to agree to the construction of hospitals *de désencombrements* (literally de-cluttering) where sick Parisians would be isolated from families and neighbors while deprived of adequate medical supervision. Care for chronic patients, or those deemed as such, the editorial argued, must take place within hospitals, close to the acute-care services they would inevitably need as their illnesses evolved.[40]

Grivaux, Huet, and Péquinot were not the only medical writers who saw chronic disease as more than a problem of patient clutter. During the 1950s, an unusual medical figure, Pierre Delore of Lyon, took up the issue in a number of publications. Delore became well known during the 1930s and 1940s for his leadership of the Neo-Hippocratic movement in France that proposed a medicine of "synthesis," or what we now call holism, to replace the scientific reductionism of modern medicine.[41] He eventually became professor of hydrology and climatology at the Medical Faculty of Lyon. He was thus active in several domains where chronic disease was a relevant concept: traditional holistic medicine of individual constitutions affected by environment; mineral waters dealing largely with management of chronic conditions; and the world of Lyon social Catholicism that had a long tradition of care for the elderly and incurable. In 1951 he spoke at the International Congress on Hospitals in Brussels that had as its theme chronic and elderly patients and at which E. M Bluestone, former director of the Montefiore Hospital in New York City, delivered the keynote address.[42] Delore's message was that the mainly elderly chronic patients in hospices were not getting the active therapy they needed and were unnecessarily deteriorating both physically and mentally.

Delore returned to the subject in the late 1950s, citing several American examples and names. While he contended that the problem was one of older people, who were becoming a larger segment of the hospital population, his message applied to younger chronic patients as well: active treatment should replace mere lodging (*hébergement*). In the name of "humanization" and "rationalization" he called for rethinking the notion of chronicity. His suggestions were not very radical. Hospitalization should be avoided and people kept at home as long as possible through hospital-based homecare. But where hospital care was needed, it was vital to avoid the social exclusion, stigma of poverty and welfare, and depersonalization that surrounded chronic disease and old age. Delore deplored the segregation of the elderly sick, who he thought should remain in close proximity to younger patients, but agreed that hospitals should have a geriatric service for purposes of education and research. His major criticisms were directed at hospices, whose existence was based on the assumption that sickness in old age was incurable. On the contrary, he insisted, suitable long-term hospital treatment could ameliorate or even cure many elderly patients who were now in either hospices or acute-care hospital wards where, in both cases, they received inappropriate care. The solution was to transform hospices. Delore also suggested that the terms *hospice* and *chronic* needed to be changed because they had negative, even stigmatizing, associations. Once hospices provided adequate treatment, they could be called hospitals for long-term illness. His program was based on optimism that rehabilitation could substantially improve the health of chronic and elderly patients: "one must always try something before announcing that there is nothing more to be done."[43]

In a later book, American influences were clearer. Delore disagreed with the American tendency to define chronic illness as anything longer than three months, proposing instead that chronic diseases were conditions that did not heal spontaneously and that with time could become irreversible. He now replaced the term *chronic illness* with *maladie a évolution prolongée*,[44] to indicate that chronic illnesses started at a young age and developed over a long period. The physician thus had to intervene among younger adults and even children in order to modify unhealthy behaviors of every kind. The American insistence on self-responsibility was firmly embraced: "health is conserved by a daily discipline, it implies an intervention by the individual himself and not just by the collectivity in dealing with the illness."[45] Other strategies included periodic medical examinations and effectively confronting environmental dangers of all sorts. When illness appeared, its progress could be arrested and improvement effected through techniques, like physiotherapy, kinesiotherapy, occupational therapy, psychotherapy, and diet. (Delore cited New York City's famous rehabilitation center at Bellevue Hospital as a model of such efficacy.)[46] He had less faith in medications (with the typically French exception of hormones), which he believed had only symptomatic effects.

It was vital to keep institutionalized patients active, and avoid the standard practice of keeping them bedridden. His textbook cases were the various forms of chronic rheumatism that comprised about 10% of all invalidity. In another prescient moment he suggested that geriatrics could become "the great refuge of general medicine," a field that in France as everywhere else was shrinking due to the rise of specialties.[47]

If Delore was an unusual figure in the French medical context, so was a second individual advocating new chronic disease strategies. Henri Péquignot was, like Delore, a practicing Catholic with holistic inclinations. Before going on to a career as a hospital physician and faculty professor, he had worked in the Ministry of Health during the early 1950s where he was a leading figure in the development of the field of medical economics, then mainly concerned with the practical planning needs of the healthcare system.[48] In 1955 he was lead author of a study of patients in several hospices in the Paris region. The study found that their age of entry had gone up significantly (which might have been the result of long waiting lists) but that these patients were also living to a significantly greater age. Péquignot concluded by pointing out almost in passing that about 20% of these elderly inmates were in hospice infirmaries where three-quarters stayed longer than three months. He concluded that elder hospices were becoming effectively institutions for the chronically ill. Equipment and personnel needed to adapt to this reality.[49] In a survey conducted several years later, Péquignot concluded that rather than getting hospital care they did not need (as medical orthodoxy claimed), elderly chronic patients were seriously underserved medically. Pointing to studies by the British gerontologist R. E. Tunbridge showing that people over 60 seriously overestimated their health status, Péquignot suggested that the elderly population had spent most of their lives before curative medicine showed effectiveness and thus had not developed the habit of turning to medicine when not well. This was certainly not the case of the following generation, so that doctors and administrators had to be prepared for a future elderly population that was much more demanding of medical care.

By 1960, Péquignot had a fully articulated conception of chronic illness that went directly against the widespread notion that scarce hospital resources needed to be rationed in favor of younger acute patients. In his view, treating chronic disease, primarily of elderly people, had now become the principal function of medicine and of hospitals and it was necessary to adjust to this new reality. Segregating chronic patients in separate institutions or wards was to be avoided unless these provided superior services to those available in general hospitals.[50] The latter were ordinarily the only institutions with the technical resources to deal with elderly chronically ill patients, many with multiple conditions requiring a variety of medical services. Like McKeown, Péquignot thought that it was important for every doctor, nurse, and medical student to spend some time with chronic patients in order to understand the limits of medical science and because this had become the central medical problem of the age.[51] Thus, constructing more hospices or lightly medicalized chronic-care hospitals, as administrators had first attempted to do, or creating a few special geriatric services as a few more enlightened figures were now proposing, were not acceptable strategies. Half of all hospital facilities should be devoted to older chronic patients who required more rather than less medical personnel and resources than younger patients.

In the winter of 1964–65, Péquignot brought his concerns to the prestigious Paris Academy of Medicine, where he presented two talks arguing that chronic disease, having now become the major source of morbidity, should not be segregated from acute care. Every hospital ward should have integrated chronic-care wards.[52] Academicians listened respectfully but few responded favorably. Pierre Damelon, a lifelong bureaucrat and now head of the Parisian hospital administration, responded that the system was already satisfactory. As a result of the hospital reforms of 1958, university and regional hospital centers made up the first tier of quality care. Then came second-level hospitals, which included chronic-care hospitals in which patients enjoyed reasonable levels of physical comfort and good medical and rehabilitation services. Patients who recovered or achieved functional autonomy were sent home. Those who did not (or had no home to go back to) were sent to a third category of institution, hospices. These, Damelon admitted, remained problematic and would be dealt with by the administration once the future of hospitals was assured.[53]

Damelon's predecessor as director of the AP, Xavier Leclainche, was more blunt. He characterized Péquignot's views as "certainly original and in this respect rather surprising." This was not meant as a complement. He refused to accept that "the needs of a chronic are, not only equivalent, but superior to those of other maladies." The complexity and multiplicity of illnesses among the elderly population did not in his view justify utilizing more diagnostic and therapeutic services for them than for acute cases. He, moreover, saw nothing wrong with special hospitals geared to the medical, rehabilitation, and leisure needs of chronic patients. He admitted that the existence of so many hospices without adequate medical services was a grave problem, but the administration was committed to dealing with them. The bottom line, however, remained the traditional Malthusian argument: given the continuing insufficiency of hospital resources, "I believe that it is more reasonable to reserve our best resources for treating the 'acutely' ill (*des malades 'aigus'*), that is to say the sick that are in general younger, suffering from recent conditions that are more completely curable, whose threatened lives can be saved more rapidly and more surely."[54] He concluded that the plan to integrate chronic sections in each hospital service was unacceptable and made little sense from a theoretical or practical perspective. One would have to reverse decades of hospital policy.

It was finally decided to create a commission to discuss the issue and report back to the academy. No report ever appeared, however, and the problem largely disappeared from the public arena. The issue of chronicity did not go away but was dealt with using a rather different institutional logic and vocabulary than the one proposed by Péquignot or by American reformers. To put it simply, the problem of chronic illness, never a major social priority except as a nuisance for hospital administrators and physicians, was absorbed into two broad issues that largely overshadowed it: the general problem of France's growing older population and the ongoing reform and rationalization of the French hospital system.

The French Crisis of Aging

Chronic disease, for Delore, Péquignot, and other French writers was essentially about the health status of a sick or infirm minority of older people. In this respect, it was similar to British unease about elderly people who remained inadequately treated. But because France lacked a strong tradition of epidemiology or a charismatic figure like Jerry Morris to popularize it, the American notion of chronic disease as a comprehensive problem was not part of the vision of French reformers, who focused on provision of services. Understood in this way, the illnesses of old age became during the 1950s and 1960s one element of a broader problem: reintegrating into society an elderly population living in a world of social exclusion and marginality.

During the nineteenth and early twentieth centuries, old age as a social category was identified with indigence and inability to work. During the interwar and post–World War II years, the spread of retirement and pensions expanded the meaning of old age and transformed it into a socially heterogeneous population group, as Anne-Marie Guillemard's magisterial studies have demonstrated.[55] Although the political problem of the aged in France remained secondary to the need to expand population through family policies, it nonetheless became a major public issue in political, administrative and media circles, with dozens of books and hundreds of articles published on the subject, including a famous book by Simone de Beauvoir.[56] Henri Péquignot wrote three books on the topic (actually updates or variations of his major themes), the last two graced with a preface by the influential philosopher Georges Canguilhem.[57] The problem was not so much the growth of the older population as whole, but of very elderly individuals who were less able to care for themselves. The number of people over 85 rose from 200,000 in 1950 to 700,000 in 1985 with projections that it would go beyond a million by 2000. Surveys showed that as many as 15% of those over 75 suffered from serious physical or mental incapacity and/or lived in an institutional setting.[58]

During the 1950s, Guillemard has shown, old age remained associated with poverty so that policy revolved around increasing or adding new stipends and forms of assistance to those assumed to have inadequate incomes. Starting around 1960, the situation changed at least at the rhetorical level. The problem of older people, it was determined, went beyond the poor and extended to the entire aged population. (Some would say that it became essentially a problem of the middle-class elderly.) And the chief difficulty was social isolation, a condition that could only be remedied by active efforts at social "reinsertion." The turning point was a commission appointed by the prime minister in 1960 and chaired by Pierre Laroque, considered the founder of the postwar social security system in France. This commission produced a long and influential report in 1962 that moved the center of policy concern for the elderly population from poverty toward quality of life and particularly their place in the community. This meant keeping older people autonomous and at home for as long as possible using a variety of social services, including adapted housing, domestic help, social services, and leisure activities. Certainly part of this strategy was to keep people out of hospitals and hospices through adequate healthcare. But the Laroque report, a volume of more than four hundred pages, included only a single chapter of twelve pages on the health needs of elderly citizens;[59] regular medical examinations, hygienic education, homecare, and ambulatory treatment were part of the attempt to keep older individuals integrated within their communities. Some old people required hospital care but despite general admiration for British healthcare policies, the report was generally opposed to segregated geriatric wards, except in a few teaching hospitals. The report did propose much wider access to functional rehabilitation services. But on the whole, such care was only one aspect of a policy that focused primarily on the quality of life of the elderly population and that left little room for targeting medical problems let alone chronic illness. A medical geriatric approach was developing in parallel to this social approach. In 1961 the French

Geriatric Society was formed, and in 1970 an official report by François Bourlière, a leading geriatrician, outlined a medical response to aging. This suggested the creation of comprehensive geriatric units in university hospital centers, a measure that was instituted on a limited basis. Although the two reports underscored the sharp separation between social and medical responses to an aging population, it has been pointed out that the public outcry against the "exclusion" of older persons provided the space for geriatrics to grow as a medical specialty.[60] Nonetheless, despite enormous political concern about an older population, geriatrics grew much more slowly in France than in the United Kingdom. Although postgraduate training and diplomas were available in the field, it did not become a full-fledged specialty until 2004. As late as 2011 geriatrics did not appear on lists of specialties although nearly three thousand general practitioners (out of a little more than one hundred thousand) were listed as having a competence in geriatrics.[61]

Not only was old age in France no longer associated with poverty or illness; it was identified as a new stage of life, the third age (*troisième age*), a term that was coined around 1955. Despite much initial rhetoric, the Laroque report's recommendations only began to be implemented at the end of the 1960s, when officials began putting together the Sixth Plan for Economic Development (1971–75) that made provision for elderly people a major priority and that allocated over five years 1 billion francs for this purpose.[62] Social marginality and reinsertion, in fact, was something of a framework for a variety of social problems during the 1960s and 1970s, where "reinserting" the "excluded"—whether ill, disabled, or otherwise disadvantaged—back into the community became the new policy orthodoxy. By the 1970s, "exclusion" and "reinsertion" were formulas applied to old age, mental illness, physical and mental handicap, immigration, unemployment, and poverty.[63] This usage reflected hostility to institutionalization but also a new ideological perspective that reformulated issues of social inequality. After 1980, while ideas of community integration retained their policy significance for older people and the handicapped, the concept of "exclusion" became chiefly identified with poverty and achieved immense influence in social science and policy circles.[64] Within both the old and new frameworks, chronic illness was not a relevant category.

Hospital Reform

Despite the new public interest in older people, the focus of health policy from the 1970s on remained the nation's hospital system. The expansion of the health insurance system, the Sécurité Sociale, throughout the 1950s and 1960s led to significant state and local investment in new hospitals, treating an expanding population with increasingly expensive technologies. Uncoordinated growth resulted in wide variation in facilities, with some regions having too few and others too many institutions. It also resulted in great expense; hospital costs made up 40% of the health budget and grew at twice the rate of the gross national product. Rationalizing the system under state control came to be seen as the way of stemming costs and adjusting supply to demand. Three pieces of legislation, in 1970, 1975, and 1978 (along with numerous administrative circulars and decrees), attempted to restructure the hospital system as a "public service" and end what was termed its "balkanization."[65] At stake were several issues: the power of the central government versus that of local authorities, financial reform, efficient geographical distribution of hospital facilities (*carte sanitaire*), and coordination of institutions, including private hospitals, to allow for rapid movement of patients to appropriate settings as their conditions changed. At the heart of these reforms was the sharp differentiation between institutions defined as medical and those defined as social or medico-social. Of particular interest for our purposes was the conscious decision to eliminate the terms *hospice* and *chronic* from the policy vocabulary (as Delore had advocated years before). To some degree this reflected the popular association of these terms with indigence, medical abandonment, and misery. But it also expressed a new administrative logic that replaced old juridical definitions and subjective judgments about patients' prognosis by strictly functional and quantifiable criteria, namely length of stay.[66] These reforms continued earlier efforts to move the chronically ill out of urban acute-care hospital wards.

Length of illness was not a new criterion and was already central to health insurance reimbursement policy. A category of illnesses considered long-term (*maladies de long durée*, later *affections de long durée*) had for many years been exempted from user fees demanded for most medical procedures. Now this logic was elaborated and extended to the hospital system. The law of 1970 established a distinction between short-term, middle-term, and long-term care. Entire institutions as well as different services and wards in the same hospital could be classed according to these categories regulated by state norms.[67] Short-term meant acute-care interventions for patients with urgent needs. There were several classes of short-term hospitals, depending on size, services, and geographic jurisdiction. Middle-term care was the most complicated, including all those who, having received acute-care hospital treatment, required more prolonged care for purposes of convalescence, rehabilitation, or simply because their condition took longer to resolve. These different rationales for treatment, initially jumbled together, were by 1980 recognized as distinct subcategories of middle-term care. The three chronic-care hospitals in Draveil, Villiers-le-Bel, and Sevran constructed during the 1960s were placed in this middle category, as were rehabilitation centers. While middle-term (later to be called "follow-up") institutions were originally meant only for patients previously hospitalized in acute-care wards, this requirement was soon dropped. By 1997, administrators were complaining that categories were not clear and that institutions contained many patients unsuited for the care offered.[68]

Long-term care was geared for people who could no longer live autonomously, which largely meant elderly and severely disabled individuals. In 1973 there were about 164,000 places in 872 long-term hospital units and a smaller number of hospices under the administration of hospitals. The Paris AP, for instance, had 13 such institutions in 1976 with a total capacity of more than 6,500 beds, including one in Ivry with nearly 1,700 beds and the Paul Brousse Hospital in Villejuif with more than 1,000. These cared for long-term patients with significant medical needs. The number of such beds had declined by about 45% between 1966 and 1976, as many were transferred to social welfare institutions that provided significantly less medical care. Laws passed in 1975 and 1978 completed the transfer of non-hospital-based hospices (now called "retirement homes") from the medical administration responsible for hospitals to a separate "social" administration where they were associated with institutions for handicapped children and sheltered workshops. Among the goals of reform was to modernize these institutions and make them smaller, more patient-friendly, and more easily accessible.[69] Because hospitals were allowed to maintain up to two hundred long-term beds, two distinct categories of long-term care were thus created, one attached to hospitals and one not. The former were under the control of a "sanitary" (medical) and national administration while the latter were under "social" local administration.

Application of these laws was messy. Although most hospices were supposed to be converted into retirement homes, some applied for long-term care status that would allow them to accept patients requiring more intense care (and that would provide better reimbursement). A special committee to decide on the category to be attributed to each institution began to function in 1981. By 1983, eighty thousand hospice beds were categorized, with about 85% assigned to retirement homes and 15% to long-term care. At the end of 1986, more than one-third of hospice beds had still not been classified.[70] The most virulently denounced aspect of these reforms was the isolation of long-term institutions administered by "social" authorities. This distinction assumed less medical need on the part of elderly people, who constituted about 85% of the retirement home population. It was assumed that most of them were healthy and had minimal medical needs so the goal was to improve quality of life in line with the exclusion/insertion framework. It was nonetheless recognized that many elderly clients also had significant health problems. From the early 1980s on, medical care sections dealing with health issues were created in certain hospice institutions. These cost about half of what long-term hospital beds cost and more and more such beds were created during the 1980s. Although regulations limited these *sections de cure médicale* to one-quarter of total institutional capacity, this quota was quickly surpassed. These reforms created considerable confusion, conflict, and ambiguity.[71] In an attempt to narrow the gap between hospitals and former hospices, a reform law of 1991 allowed hospitals to manage "social" as well as medical institutions.

One could say much more about hospital reform, which has been a constant struggle to bring greater rationalization to a huge and unwieldy system that does not yield easily to rationalization. There have been nine major hospital reforms in France since the start of the Fifth Republic in 1958.[72] The new organizational logic based on length of care was not just a matter of changing labels and making the word *chronic* an informal or folk term rather than a policy category; it had practical implications. The focus of healthcare policy in France has been on the provision of a *system* of institutional facilities, particularly hospitals. During the 1980s, there was a significant rise in capacity of mid-length units in the public and nonprofit private sectors.[73] Long-term care institutions existed in large numbers, with the private sector becoming increasingly involved in what proved to be a lucrative trade. The administrative and physical separation between most former hospices and the hospital system with its own long-term care wards proved to be the most controversial aspects of the new arrangements. The problem, in fact, worsened as patients were admitted to retirement homes at an increasingly advanced age and frequently requiring medical supervision. Consequently, institutional care for older people has taken on an increasingly medical character.[74] But more than medical care has been involved. As the old lived even longer, giving rise to a new term, the *fourth age*, a new category appeared in the policy vocabulary: "dependence." Referring to individuals too infirm to live autonomously and requiring help for the tasks of daily living, "dependence" has since the late 1970s become one of the chief preoccupations of French policy for older people. Curiously, *dependence* is a term reserved for very old people in France and is carefully distinguished from services needed by younger adults unable to live autonomously.[75]

Aside from separating social from medical care in the case of older people, the French have also made another distinction that complicates policy. Disabilities, or what the French have come to call "handicaps," are sharply distinguished from "evolving diseases," chronic illnesses that grow worse with time and that frequently lead to serious disabilities. From the earliest days of the Sécurité Sociale, long-term illness was distinguished from disability within the insurance and entitlement systems. A law passed in 1975 extending entitlements to the handicapped without defining them gave rise to a commission, sponsored by the major national center specializing in handicap research, to study the relationship between handicaps and evolving diseases. Its report, published in 1979, resoundingly rejected the standard justification for this distinction—that disabilities or handicaps are stable and permanent while illnesses constantly evolve. Handicaps not only evolve with time, it was argued, but their significance is relative to social circumstances (losing a finger is quite different for a violinist than for a historian). The commission came up with a plan bringing together the two categories but also providing comprehensive proposals for dealing with them that included prevention, treatment, family support and education, and vocational rehabilitation. The plan was in fact the closest thing to the comprehensive American vision of chronic disease, albeit using different terms, produced in the French context. It, however, went nowhere. As its authors admitted, sick people and their doctors (notably mentally ill patients and psychiatrists) rejected classification as handicapped, and after 1980 the handicapped increasingly rejected the identification of their conditions with disease.[76]

While hospitals have played a much greater role in chronic care in France than they do in the United States, which relies more heavily on private nursing homes, this has not precluded development of non-institutional care. The policy developed by the Laroque Commission to maintain elderly people within the community was predicated on the development of homecare. Policies around "dependence" have continued the effort to keep very old individuals out of institutions for as long as possible. The state policy on this matter has been seriously criticized for being underfinanced, too much dominated by geriatric medicine, and focused on creating jobs for unskilled women.[77] Nonetheless, by 1990, about 500,000 people were getting domestic home help. The number of places in assisted living facilities rose from under 10,000 in 1968 to more than 142,000 in 1990. Homecare was available to 11,700 people in 1982, 45,000 in 1992, and nearly

72,000 in 2003.[78] Special financial allocations to permit autonomous living were introduced during the 1990s, and by 2003, 600,000 people were receiving them. Highly specialized hospital-organized homecare, called L'hospitalisation a Domicile (HAD), has had less success. It was introduced in Paris on a small scale in the late 1950s and grew slowly. In 1973 there were ten units in existence nationally and in that year a federation of such programs was created. By 1982, the number of programs had risen to twenty-five, mainly in Paris and other large cities. Like homecare funded by Medicare in the United States, HAD was meant to follow hospitalization. The situation persisted until 1986, when HAD was allowed to be prescribed medically without prior hospitalization. Nonetheless, the sector stagnated. A law of 2003 proposed to double the capacity of HAD programs in order to serve eight thousand people annually by 2005.[79] Recent work suggests that considerable development has occurred since, with most emphasis on the areas of palliative care, cancer treatment, and perinatal care.[80]

Public Health and Prevention

I have so far said little about public health and prevention in twentieth-century France. In part, this reflects the relative lack of information about the subject, which itself mirrors the surprising weakness of public health institutions in this highly bureaucratized nation.[81] Greater investments were made during the post–World War II period but these, it has been argued, were overshadowed by an overwhelming policy focus on healthcare institutions and biomedical research.[82] The Sixth Economic Plan (1971–75) emphasized programs of prevention and screening, but when the plan came to its end, it was concluded that, except for improvements in highway security and perinatal screening, the situation was little changed.[83] Overall, the WHO determined in its 2000 health report that while France was doing well in the medical care field, it did not boast very good results in the field of prevention; mortality avoidable by proper prevention was far too high considering the nation's economic and educational status.[84]

Equally significant for our purposes is that the most relevant category of public health work in France was not "chronic" or "acute" disease, but "social diseases" (*maladies sociales*), often referred to as social plagues (*fléau*), referring to diseases that affected large numbers of people and/or had some kind of social origin and thus required administrative intervention. Tuberculosis was the original social disease. The Institut National d'Hygiène (INH), created by the Vichy regime as a public health research organization in 1941, had a section devoted to social disease that focused on tuberculosis, syphilis, alcoholism, and cancer. After the war, the number of social diseases increased.[85] At the level of public health practice, the section of Social Hygiene in the Ministry of Health was advised by a permanent council with nine committees covering cancer, sanitary education, leprosy, venereal disease, mental illness, maternal health, child health, and rheumatism, in addition to the omnipresent tuberculosis. In 1955 commissions were added for diabetes, old age, and infrastructure. Many of the conditions being dealt with were what Americans called chronic diseases but the logic was that of discrete and wildly different problems, each to be dealt with in theory through comprehensive multidisciplinary approaches to prevention, treatment, and rehabilitation. (In practice multidisciplinarity was limited.)[86] The subsection changed names several times before being divided in 1976 into three different administrations: prevention and detection, treatment and rehabilitation, and maternal and child health.[87] The Sixth Economic Plan for the period 1971–75 also followed this logic, listing seven priority problems, including traditional ones like cancer, perinatal mortality, and degenerative diseases of old age, as well as newer ones like accidents and cardiovascular disease.[88]

The INH was transformed in 1964 to become INSERM (Institut National de la Santé et de la Recherche Médicale), a research institution that gradually took on a predominantly biomedical orientation and that deemphasized public health research. Rather than chronic disease epidemiology, what emerged in France was a new kind of biomedical statistics pioneered by Daniel Schwartz, who became the dominant figure in the field. Epidemiology in this new form was essentially a service discipline. It focused on three areas: case control studies, starting with the effects of smoking and going on to many other disease causes; clinical trials; and laboratory experimentation. The first of these was perfectly congruent (and in close contact) with the risk factor epidemiology that had developed in the United States and Britain. But its practical effects were limited because of the small number of Schwartz-trained or -inspired epidemiologists, most of whom were not physicians; the absence of a strong public health establishment to back them; their narrow technical focus and overall lack of interest in public health issues (there were no theoreticians like Jerry Morris to link epidemiology to public policy); and epidemiology's relatively low status in the hierarchy of biomedical research. Although its practitioners worked on the same diseases as their English-speaking counterparts, they were not a force promoting a comprehensive understanding of chronic disease.[89]

One of the reasons that public health was not well developed is that so much was demanded of hospitals. According to the hospital law of 1970, hospitals were expected to take a leading role in prevention and health education. It was soon generally agreed that they were not up to this task. Prevention and education, according to the report by Paulette Hofman for the Economic and Social Council in 1983, was "one of the domains where the public hospital service is far, very far even, from having gone all the way toward fulfilling its possibilities."[90] The single exceptions, she conceded, were some programs in maternity and early childhood prevention. That hospitals failed to become centers of prevention is hardly surprising given their evolution toward ever-greater technological sophistication and commitment to rapid turnover of patients. That the French relied on hospitals for so long speaks to the central role these institutions were seen to play in that nation. While there were and continue to be numerous administrative agencies responsible for older and/or handicapped individuals, there emerged few if any administrative structures directed at chronically ill people more generally. The focus continued to be on providing access to hospital care and shortening hospital stays. Previous failures did not prevent the WHO in 1990 from launching a new Health Promoting Hospital Program to aid hospitals in implementing health promotion interventions. In France one year later, a law reforming hospitals again included among their missions prevention and public health but without specifying how or with what resources this should be done.[91]

The French, like the British, were cautious about public health measures that did not have proven benefits. They seem to have ignored international excitement surrounding multiphasic screening during the 1960s and 1970s. One reason was the focus on hospital reform and another was the strict conditions for establishing programs set out by the WHO-Wilson-Jungner guidelines. But long before these guidelines were published, the ordinances setting up the social security system in 1945 introduced periodic physical examinations. Anyone with insurance was given the right to a free physical examination once every five years. Insurance funds set up special examination clinics to take blood, x-rays of lungs, and so on, with tests followed by a physical examination. Initially, the main diseases targeted were tuberculosis and syphilis but this changed as the incidence of these illnesses declined. Only a minority of the population, roughly 650,000 adults annually in the late twentieth century, used the service, which was available in 98 centers. The practice was vigorously criticized, especially following the report of the Canadian Task Force on the Periodic Health Examination in 1979. A reform in 1992 oriented this examination to populations that were socially marginal or "precarious," presumably more vulnerable to illness. Free health examination remains a contested practice in France as in much of Europe because it affects so few people, does not have a sharp diagnostic or screening focus, and shows no clear evidence of improving the health of those receiving it.[92] It is not surprising that expanding examinations to the entire population initially had little support in France. The commission supervising national economic planning included "check-ups" among the fashionable "crazes" (*engouements*) that should be avoided.[93]

As in other countries, media public health campaigns became a key method of preventive work. Although there had been interesting efforts during the interwar years to use film for health education,[94] modern forms of media education were developed somewhat later in France than in the United States or Britain. Although an agency to promote "sanitary education" was created in 1945, it was not until 1972 that an organizationally stable and well-financed agency, the Comité Français d'Education pour la Santé, was established. The government of Giscard d'Estaing then made a major commitment to "modern" mediatized healthy lifestyle campaigns, starting with the 1976 campaign on the risks of smoking. The agency charged with such activities organized eleven more public health campaigns, including ones centered on the risks of a sedentary lifestyle (1977), abuse of medications (1980), and excessive alcohol consumption (1984), the latter being the first such campaign to use television extensively. As Luc Berlivet has noted, these campaigns became increasingly sophisticated as their organizers learned that information was far less important than transforming motivation.[95]

The first national programs of targeted screenings in France for all neonates were established in the 1970s for phenylketonuria and congenital hypothyroidism.[96] Although the benefits of immediate treatment following cervical and breast cancer screening were not so clear-cut, these became increasingly popular in the 1980s within the context of private medical practice and reimbursement by the health insurance system. Such informal screening was criticized for a variety of reasons. The wrong people (young women) were getting too much of it (annually) and those at risk (women over 50) were not getting enough. And there was no way to really monitor or evaluate screening in such contexts. The issue was less pressing in the case of cervical cancer screening, which was very popular among French women; public health activity mainly centered on encouraging those who had never been tested to do so during doctor visits. In 1990 a number of local pilot programs were undertaken and all cervical screening tests made free.[97] Breast cancer screening, however, proved more complicated.

In 1986 the European Union initiated a new program, "Europe Against Cancer," to finance pilot programs of cancer screening. The man who spearheaded this new initiative was the French cancer specialist Maurice Tubiana. In 1988 a private association made up mainly of physicians was founded in Strasbourg and received European and French financing to develop a breast cancer screening program in the surrounding department of the Bas-Rhin. In 1989 one of France's major health insurance funds covering salaried workers, the Caisse nationale de l'assurance maladie des travailleurs salariés (CNAMTS), introduced its own pilot projects in ten departments. These programs created the French model for screening; financed nationally, they were administered at the local level and used existing medical resources, notably the large number of radiologists in private practice. In 1993 the minister of health decided on a national anticancer program with screening among its major components. A year later, a national committee under the presidency of Maurice Tubiana was created to manage it. Tubiana could write in 1997: "Practically ignored twenty years ago, mass screening has today become one of the magic words that all political and medical authorities must pronounce."[98] By then, twenty-four departments had organized pilot screening programs. That this constituted less than one-quarter of all French departments was disappointing, as was the less than 50% participation rate of women; these numbers were explained by the fact that so much mammography, whose effects were impossible to evaluate, was going on outside of the formal programs. (One study concluded that France had 10 times more mammography machines than did the United Kingdom with its already functioning national screening program.) Consequently, one of the goals of the national committee was to harmonize spontaneous with organized screening. To these barriers slowing down the spread of national programs could be added inadequate financing, lack of national direction and cooperation from professionals, and much bureaucratic red tape. In 2003 many departments still lacked programs, despite regular ministerial assurances that breast cancer screening would soon become fully national. The participation rate of women in these programs remained disappointingly low, even as rates of private individual screening continued to rise.[99] In 2004 the breast cancer screening program was finally extended to all French territory. An earlier ministerial announcement to this effect, also indicating that screening for uterine and colon cancer would be considered in the near future, prompted one medical journal to title an article on the subject: "Cancer Screening: The End of French Backwardness?"[100]

One area that seems to have been relatively successful was rehabilitation. Rehabilitation centers developed in France's rationalized healthcare system in a somewhat more coordinated way than in the United States thanks to centralized planning. A very small and marginal field, physical medicine and readaptation (PMR), began to develop during the 1950s because of political pressure to do something about the sequelae of poliomyelitis. PMR departments were created in a variety of hospitals, including the chronic-care hospital at Garches. Many sanatoria were transformed into public or private rehabilitation centers and new institutions were constructed. A national commission on rehabilitation developed a comprehensive plan that was turned into policy by a government circular of 1960. In 1973 PMR was recognized as a medical specialty with full national certification while the hospital reforms of 1970s recognized rehabilitation centers and units among specialized middle-term hospital units (with appropriate reimbursement). By 1997, there were 1,780 specialists in this field with almost 300 of them working in hospitals. By the end of the twentieth century, 18 research laboratories devoted to rehabilitation were in existence. For all the complaints that its status is low and the number of specialists insufficient, the field seems relatively well developed in France.[101]

The existence of rehabilitation has been somewhat complicated by official interest in yet another group characterized by marginality and exclusion: handicapped people. As with old age, a national report in 1969 crystallized official recognition of the problem.[102] A law passed in 1975 conceived within the "exclusion/insertion" framework, enshrined the legal rights of handicapped individuals, based on their entitlement to insertion into normal society. The first article of the law made it a national obligation to train and orient them professionally while providing them with the guarantee of sufficient resources, but also (and these were among the first words of this article) to provide prevention, screening, and care.[103] Nonetheless, growing identification of handicap with "exclusion and marginality" since the 1980s has transformed it into what is widely understood as a social problem. This, it is suggested, has made rehabilitation a less visibly medical activity, resulting in its absence from healthcare planning. According to this view, it is almost exclusively seen as a response to existing functional problems. Conflicts among institutions and interest groups, it is claimed, have exacerbated this situation.[104]

Conclusion

By the late twentieth century, French medicine had many of the trappings associated with American chronic disease medicine: a network of institutions that cared for chronic disease patients of all ages (and, in addition, a national insurance system that paid for their care); various forms of homecare and long-term institutions, primarily in the public sector but with a growing number in the private sector; a small but carefully chosen number of national screening programs and mediatized campaigns promoting healthy behaviors based on risk factor epidemiology; and a system of rehabilitation facilities integrated with the hospital system. And all of this came about without invoking the menacing specter of chronic disease. Two factors were critical to this process. The first was the decision to create a national health insurance system that could serve as the basis for healthcare planning for the entire population, not all of which suffered from chronic conditions; the second was the strenuous effort to rationalize the nation's system of hospitals and long-term care through which patients could move at various stages of life and disease. There seems to have been no need to mobilize around the threat of a chronic disease crisis. At least that was the case until the early twenty-first century, when the French quite abruptly discovered the utility of conceptualizing policy in terms of chronic disease.

Epilogue

For much of the twentieth century, France and the United Kingdom faced somewhat different health issues than did the United States. Their populations were older, although by the 1970s the age gap had narrowed and Medicare had made the health of older people a significant part of the American chronic disease problem. In both European nations, illnesses of younger and middle-aged persons were treated within the context of national healthcare systems and were sharply distinguished from those of older people, widely viewed as "bed-blockers" and as a problem left whenever possible to social agencies. The acute medical needs of the elderly could be met by hospitals and persistent ailments handled by special institutions and the specialty of geriatrics. In contrast, Medicare in the United States took a narrowly medical approach to reimbursable services for older people, reflecting the predominant view of chronicity that had emerged in the 1950s. Strikingly, "chronic disease" did not play the significant policy role in France and Britain that it played in the United States.

This situation has recently changed—in the case of France quite abruptly. Since about 2000, "chronic disease" has emerged in both France and the United Kingdom as a public issue that has generated new policy initiatives. To a large degree this is a response to problems internal to their healthcare systems, but it also reflects policy discourses that are circulating within a wide range of international institutions like the World Health Organization (WHO) and the Organisation for Economic Co-operation and Development (OECD), not to mention regional groupings like the European Union and private associations like the European Public Health Association, which established a chronic disease section in 2006. As in the 1950s and 1960s, American models seem to be the source of much of this European rhetoric and activity. But each country (and each organization as well) has put its own spin on these ideas.

Somewhere around 1980, chronic diseases began to be perceived as a problem of the developing world. Research was beginning to show that hypertension, cancers, heart and respiratory diseases, and especially diabetes, were emerging as health problems worldwide. By 1984, sixteen countries in the WHO Western Pacific Region (which encompasses East and Southeast Asia) were reporting a greater incidence of deaths from noncommunicable diseases than from infectious and parasitic diseases combined.[1] "Chronic disease" was added to international health agendas (in the form of "noncommunicable disease") through the INTERHEALTH program launched by the WHO in 1986. INTERHEALTH was about transferring the risk factor epidemiology/lifestyle-based intervention model to developmental settings. In line with the WHO's traditional emphasis on building community health services, the aim was to construct "horizontally integrated" programming (as opposed to disease-specific, "vertically integrated" programming like malaria eradication); risk factors provided a basis for such comprehensive services. The first countries to set up demonstration projects included developed nations like Finland and the United States, but also less affluent countries like Sri Lanka, Tanzania, and Thailand.[2] The WHO launched similar programs for Europe and Canada and, in collaboration with the Pan-American Health Organization, for the Americas. Intervention programs were introduced that incorporated mass media, patient education, professional training, and community infrastructure development, tailored to local needs and conditions. "Integration" was the slogan of the day, as "coordination" had been for American chronic disease administrators in the 1960s, signifying both common approaches to risk factors for many diseases and attempts to bring together "inputs, delivery, management and organization of services related to diagnosis, treatment, care, rehabilitation and health promotion."[3] If INTERHEALTH and other WHO programs reflected traditional ideas of risk factor management, they placed greater emphasis on building self-care and self-management capacity, perceived as especially urgent in poorer countries with less-developed medical infrastructures.[4]

Meanwhile, the United States was dealing with its own chronic disease problem by embracing disease management strategies. Spurred by ongoing concern to reduce rising costs and improve quality of care, as well as by the spreading model of the patient-as-informed-health-consumer, disease management emphasized coordination across healthcare services, standardized outcome measures, and the use of evidence-based guidelines to direct long-term management of conditions and reduce the number of acute episodes.[5] Two features characterized this approach. First, appropriately trained patients were to take major responsibility for their conditions through self-management. Second, managers from outside the primary care system, viewed as irredeemably oriented toward specialized acute care, were to train patients and provide information and support.

Disease management programs took a variety of forms but were usually developed by commercial disease management firms that either sold services to or developed them collaboratively with local health agencies. The disease management industry in the United States grew from $78 million in 1997 to $1.2 billion in 2005. Most management programs were single disease–oriented and designed for high-severity conditions; the most common form they took consisted of telephone reminders or conversations with nurse consultants or case managers.[6] The Centers for Medicare and Medicaid Services (CMS), struggling with skyrocketing costs, initiated a number of demonstration projects involving more than three hundred thousand beneficiaries across thirty-five different programs. Between 1999 and 2009, the CMS sank more than $750 million into disease management programs. Medicaid agencies contracted with disease management companies in more than twenty states.[7]

Partly as a response to commercial disease management programs, criticized for being driven by cost containment rather than improved care, the Group Health Cooperative (GHC), a Seattle-based nonprofit health corporation, developed what came to be known as the chronic care model (CCM). Researchers at GHC's research arm, the MacColl Institute for Healthcare Innovation, led by Edward Wagner, an internal medicine specialist with training in public health, designed CCM based on an extensive survey of the characteristics of successful chronic disease management programs.[8] The CCM, while clearly a product of the disease management world, was intended to address what its designers saw as the major shortcomings of most such programs: excessive emphasis on cost reduction; focus on individual diseases; accepting and working around a healthcare system designed for acute care; and delegating chronic care to external case managers on the grounds that primary medical care was not up to the task.[9]

Instead, the GHC group proposed comprehensive system change that oriented American healthcare to chronic disease management. The reality was more modest: a heuristic model consisting of a list of program characteristics that could then be implemented in various permutations and combinations. The CCM model was the most systematic and successful synthesis of what had by now become a widely held set of beliefs. Part of its appeal lay in its combination of evidence-based medicine (EBM) with the popular idea that community resources, holistic perspectives, and patient-centrism are essential for chronic-care provision. That it supplied a model framework with many elements from among which policy-makers could pick and choose allowed for flexibility in implementation. The CCM insisted on comprehensive rather than disease-specific programs for some of the same reasons offered by reformers of the 1950s and 1960s; prevention and management of many different long-term conditions had common characteristics. In addition, it was now widely recognized that large numbers of chronic disease sufferers had high rates of multiple, overlapping conditions. (It has been recently estimated that 66% of American healthcare spending is directed toward the 27% of the population who suffer from multiple chronic conditions.)[10] So while disease-specific programs might continue to exist, the underlying framework and some of the actors involved would be common to all.

The CCM framework emphasized the following elements: (1) establishing linkages between health providers and community and social resources, such as exercise and education programs, self-help groups, and senior centers; (2) reorganizing healthcare systems around chronic illness through changes in clinical care, financial incentives, and reimbursement policies; (3) supporting patient self-management through education and professional services; (4) increasing emphasis on multidisciplinary teams, trained nurses, and non-physician care, under the direction of primary physicians; (5) integrating evidence-based clinical practice guidelines in the form of "reminders" to patients and care providers; (6) setting up clinical information systems utilizing the most up-to-date information technology, computerized patient records, and automated reminders.[11]

In the United States, commercial, single-disease management programs continued to be the norm. CCM-inspired projects were pursued in some locations, often in combination with a risk stratification strategy, the Kaiser Permanente risk pyramid, pioneered by the California-based healthcare group, which divided patient populations according to the type and intensity of care and support needed: for the great majority, basic support combined with self-management; for a smaller group at high risk for hospital readmission and/or with special needs, disease management; and for the 3 to 5% of patients with complex and overlapping conditions, intensive multidisciplinary care.[12] The US Bureau of Primary Health Care's community and migrant health centers took a CCM approach, as did the Indiana Chronic Disease Management Program initiated in 2003.[13] Another form that the CCM took in practice was the patient-centered medical home (PCMH), a type of physician-led team approach offering each individual preventive, acute, and chronic care services. Not a new idea, PCMH took off in 2002 when it was embraced by the specialty of family medicine. A 2010 study found ninety-five single- or multiple-practice PCMH demonstration projects whose activity ranged from explicit implementation of CCM to simple case-management practices. Many programs have adopted elements of the CCM without seeking comprehensiveness or the structural transformation of medical care.[14]

But CCM's major successes have been international. In 2000, the WHO set up a Department of Health Promotion and Noncommunicable Disease Prevention and Surveillance (HPS). Its representatives worked with Wagner to adapt, expand, and rename CCM as the innovative care for chronic conditions framework that was to be the basis for its own global programming.[15] In 2005 the HPS became the Department of Chronic Diseases and Health Promotion (CHP) reflecting this new CCM orientation. In the years that followed, the WHO produced a torrent of publications and reports on the subject, sometimes referred to as "chronic" and sometimes as "noncommunicable" disease, suggesting either confusion or disagreement about terminology. The organization managed rhetorically at least to combine the older risk factor approach to prevention with the newer disease management models. A report of 2005 insisting that all health personnel needed a new training model based on patient-centered care included supporting statements by the World Medical Association, the International Council of Nurses, the International Alliance of Patients' Organizations, and the International Pharmaceutical Federation, among other groups.[16] The WHO's European regional division has partnered with national governments and various other institutions to create the European Observatory on Health Systems and Policies with a wide mandate that includes regular updates on both the theory and national practices of chronic disease policy. The OECD has also recently entered the picture, focusing on issues of long-term care for older people.[17]

In recent years the WHO's emphasis on CCM-type programs has diminished (which may be reflected in its almost exclusive use of "noncommunicable" rather than "chronic" in its titles),[18] as attention focused on less developed countries for which the care promised by CCM seems an impossibly remote fantasy. The emphasis now appears to be on prevention and simple, more basic, and less expensive healthcare goals. Key descriptors for interventions have become "essential" and "cost effective."[19] But whatever one chooses to call them, chronic or noncommunicable diseases in the developing world have become major global health issues. The 2008–13 WHO Action Plan for the Global Strategy for the Prevention and Control of Non-communicable Diseases drew attention to the pressing need to invest in chronic disease prevention as an integral part of socioeconomic development. The creation of the Global Alliance for Chronic Diseases brought together six of the world's foremost health research agencies to collaborate in finding solutions with a particular focus on the needs of low- and middle-income countries and on low-income populations in more developed countries.[20] In 2011 health ministers from around the world met in Moscow to discuss what the *New Republic* characterized as "a serious global health crisis: the rise of non-communicable diseases around the world."[21] A few months later "a High-level Meeting on non-communicable diseases" was hosted by the General Assembly of the United Nations.[22] In January 2013 the draft for a new and more ambitious WHO action plan on noncommunicable disease was placed online.[23] Its status is not yet clear at the time of writing. Increased focus on developing nations adds new layers of complexity and political disagreement to issues that are already complicated and intractable in rich countries.

Given this intellectual and policy climate—in 2005 a new journal, *Chronic Illness* (funded by Sage but with a strong British orientation), began publication—it is not surprising that chronic disease has taken on new visibility in both the United Kingdom and France. The emphasis has been on disease management models originating in the United States (which themselves include elements borrowed from European health systems) and which were highly publicized by the WHO from around 2001 to 2005. The process has been most intensively pursued in the United Kingdom with its centralized administrative structures and a well-organized primary care sector. In addition, many people really are on their own or dependent on relatives for long-term care. The United Kingdom has ranked highest among OECD countries for firms reporting employees working part-time in order to care for relatives. In 2010 approximately 550,000 people in the United Kingdom were receiving a "carer's allowance," meaning that they were paid for regularly devoting thirty-five-plus hours weekly to care work.[24] Such carers require guidance as well. In 1999, a government white paper set out a comprehensive plan to tackle cancer, coronary heart disease and stroke, and mental illness. In 2000 the *British Medical Journal* devoted an issue to the topic of disease management, co-edited by Edward Wagner, designer of the CCM.[25] At the time, "disease management" was still a rather exotic American phenomenon, associated primarily with the for-profit health sector but this soon changed. It became part of the "modernization" of the National Health Service (NHS) that took place during the 2000s. Chronic care was recognized as a priority in the NHS Improvement Plan of 2004, drawing explicitly on CCM and the Kaiser Permanente risk pyramid.[26] A key moment occurred in 2005 when the NHS health and social care model was introduced to improve care for people with long-term conditions.[27] The basic aim is to identify users of long-term care according to need and promote patient self-management and more intensive services when necessary through teams co-ordinated by a community matron using case management approaches. (By 2012 there were 4,300 community matrons in the country, a drop from a high of about 5,300 in 2008.)[28] Other experimental programs that were introduced include the Expert Patient Programme in which trained lay people teach skills of self-management. These skills are "generic" rather than disease-specific in order to support a broad range of conditions. A 2006 white paper committed to increasing enrollment in the program from thirty thousand to one hundred thousand people by 2010. Disease-specific self-management programs also exist, as does the Evercare program in which care management by nurses is directed specifically at helping older fragile patients to remain in their own homes. In 2004 a new pay-for-performance contract was introduced for family physicians providing financial incentives to improve the quality of chronic care. It focused on 10 chronic conditions and 146 performance indicators, including such things as keeping registries of patients, doing patient reviews, and providing evidence of outcomes like control of blood pressure.[29]

Although prevention is a frequently mentioned goal of the new policies, this has received considerably less attention than management and self-management of what British policymakers call "long-term conditions," a term that supposedly focuses more directly on patient experience and includes forms of illness (such as mental illness) and disability that are not usually considered "chronic diseases."[30] In this way, the longstanding British tendency to think of chronic illness in terms of consequences and services needed rather than causes and processes has been maintained. On the other hand, the traditional focus on age seems to have been expanded in favor of a more general approach that recognizes that people from many age groups are affected by long-term conditions.[31] The NHS has thus transformed the ongoing problem of chronic disease into one of disease management with the goal of using case-finding and surveillance methods to spread the burden and responsibility for such management among patients, nurses, primary care physicians, and other service providers. Although early evaluation studies did not show significant impact on hospital admissions and found significant barriers to successful implementation, this focus has been maintained.[32] In 2012, the government introduced a radical and highly divisive reorganization of the NHS, the Health and Social Care Act of 2012 that if successfully implemented will transform British healthcare. It is too early to tell exactly where things are going, but one can affirm that long-term conditions continue to be a major preoccupation. While there is less talk of disease management, and more about measurable outcomes, one of these outcomes is enhancement of quality of life for people with long-term conditions and that is defined by specific metrics. Risk prediction and technology for remote surveillance are emphasized while patient self-management and care remains central to discourse. It is assumed that by fostering self-empowerment, self-management will inevitably improve quality of life and by reducing the number of acute episodes will lower the costs of acute care.[33]

The cumbersome French healthcare system took more time to embrace chronic disease management. This relatively generous insurance plan has run regular budget deficits that have agitated politicians and administrators for some time. During the 1990s, the government experimented with a variety of techniques to control costs: demonstration programs with general practitioners serving as gate-keepers to services; clinical practice guidelines; and networks of healthcare providers to coordinate care. Specific chronic diseases were sometimes targeted by these measures but "chronic disease" remained absent as a policy category. This changed at the end of the millennium. In 1998 the government founded the Institut de Veille Sanitaire (InVS) to keep track of the health status of the population. A year later a department of chronic disease and traumatic injuries was established within it. But the real breakthrough took place in August 2004, when two laws were passed. The Public Health Law of August 9, 2004, was a wide-ranging attempt to expand the nation's traditionally weak public health policies.[34] The InVS was given expanded powers. Also strengthened was the Institut National de Prévention et d'Education pour la Santé founded in 2002 to pursue preventive activities. The law replaced an existing national advisory body for public health with the Haut Conseil à la Santé Publique (HCSP) with a broad mandate over all aspects of public health. It was composed of six commissions, including one devoted to chronic diseases and another to health education and promotion (usually directed against chronic illnesses). The law listed one hundred objectives to be attained within five years; forty-nine of these directly concerned chronic diseases. The law also specified the elaboration of five mid-range "plans"; one was devoted to cancer, but another was devoted "to the quality of life of patients with chronic disease."

A few days later, the legislature passed a major law regarding health insurance.[35] It, too, created a powerful new institution, the Haute Autorité de Santé (HAS), to take on a variety of tasks regarding healthcare practice; these included everything having to do with practice guidelines, health-technology assessment, and institutional certification. It was also to serve as the government's major advisor on health matters. One of its functions was to deal with long-term conditions, *affections de longues durée (ALDs)*, that dispensed with usual user fees for procedures and medicines. There were thirty such conditions and an open-ended possibility of receiving dispensation for other expensive conditions. This moved HAS directly into the business of chronic disease. In addition to defining which diseases were *ALDs*, it was supposed to produce evidence-based guides for doctors and patients about dealing with each of these conditions. Dispensation from user fees would depend on following the HAS protocols. *ALDs* were a major issue not so much because of the dispensations themselves but because patients classified as such were the source of so much expense. They accounted, it was said, for 60% of all healthcare costs and the figure was rising. The health insurance law had one other critical component. It created medical gatekeepers. Every French adult was supposed to choose a physician, a *médecin traitant*, a general practitioner or specialist, who would coordinate his or her care. This was not a legal obligation but the financial advantages for doing this were significant enough that within a few years close to 90% of the population had signed up with one. This *médecin traitant* was supposed to be responsible for setting up personalized versions of the protocols produced for *ALDs*.

In the years that followed, momentum around the issue began to develop. Within months of the legislation, meetings with stakeholders were organized.[36] A number of patient groups decided in 2005 to join together to pursue common goals, forming Les Chroniques Associés made up of eight groups representing illnesses like AIDS, cancer, renal insufficiency, and multiple sclerosis. One of their main goals was to be fully recognized for purposes of disability entitlements.[37] Meanwhile, a government agency published a report in 2006 on disease management programs in Europe and North American and provided a qualified endorsement for experimenting with them in France.[38]

Finally, in 2007 a modest but official plan for chronic disease was published.[39] The focus of the plan was on improving the quality of life of chronic sufferers. (One is reminded of the emphasis during the 1960s and 1970s on improving the quality of life of the elderly population.) Three other related objectives were helping patients understand and manage their malady; placing more emphasis on prevention; and gaining better understanding of the consequences of such maladies on quality of life. (The precise wording and focus shifted over the years.) The four objectives are divided into fifteen practical and relatively modest goals. These include, among others, experimenting with provision of guidance to patients to promote self-care; allowing helpers to perform certain technical procedures usually reserved for healthcare personnel but indispensable for daily life; and offering chronic sufferers all the entitlements of the handicapped. The plan called for spending 727 million euros from 2007 to 2011. The latest evaluation of the progress of the plan appeared in 2012 with a table indicating that of the fifteen goals mentioned in the plan, six had been realized while the rest were in process (*en cours*).[40] Most of the goals involved small pilot projects in such areas as patient education, augmenting homecare, education and support for care-givers, facilitating access to handicap entitlements, and developing therapeutic living facilities (*appartement thérapeutique*). Research is being done by public agencies and projects by external researchers have been solicited and financed. In 2009 a web page full of information for chronic sufferers was created[41] and there are plans afoot to create a "portal" that will lead to all relevant websites. Health authorities have introduced a plethora of disease-specific plans independently: Plan Alzheimer 2008–12, Plan Cancer 2009–13, Plan Maladies Rares 2010–14, and a half dozen others. (All these five-year plans underline just how much the economic planning that began during the 1940s has shaped French responses to social issues.) These disease-specific plans seem to have little to do with the single general chronic disease plan but most also emphasize patient support, coordination, and self-management.[42]

Also without direct relation to the chronic disease plan, HAS experts have with the collaboration of patient groups been busy preparing guides for doctors and patients on managing *ALD* maladies. By 2012 HAS had produced sixty-two guides for doctors and forty-nine for patients. And they are now in the process of creating a second generation of more user-friendly guides.[43] Nonetheless, there seems to be little data to indicate who is using them and for what purposes. If there is clearly a weakness, aside from minimal financing, it lies in the lack of human support for the management plans being created. In its annual report published in early 2013 the Cour des Comptes, charged with auditing public institutions, devoted a chapter of its annual report to what it refers to as an "unfulfilled reform" (*une réforme inaboutie*): the introduction of the *médecin traitant* supposed to coordinate individual care. That they have not done so effectively in the Cour's opinion suggests that they are doing little to support self-care.[44] The diabetes support program, Sophia, in existence since 2008, employs nurse-counselors to provide telephone information and support to patients; it seems to have had some success and now has 226,000 diabetic members. In 2012 the major French insurance fund, CNAMTS (Caisse nationale de l'assurance maladie des travailleurs salariés), seeking to expand this program nationally, hired an international health management company, Healthways, to manage Sophia. With or without external management, many doctors are deeply hostile to the program and refuse to participate.[45]

Chronic disease is finally visible in France. Newspapers and magazines announce that the number of people classified with ALDs has risen from 7 million in 2004 to more than 9 million in 2010 and that more than 15 million people in France suffer from a chronic illness. Curiously, this last figure represents from one-fifth to one-sixth of the French population, the same proportion that the National Health Survey found in the United States in 1935–36. The Roche Pharmaceutical Company has since 2010 sponsored an annual meeting on chronic disease with the same patient-centered, quality-of-life approach taken by the 2007 plan. It is true that relatively small amounts of money are being spent on such programs and most of that is devoted to disease-based plans. But the concept of chronic disease combined with the idea of patient self-management has most definitely arrived in France.

What then is one to make of this global attention to chronic illness? The issue in developing countries is far too complex to discuss in a brief epilogue and in any case this book deals with three affluent countries. The conventional and in some ways correct explanation for the growing clamor about chronic, noncommunicable, or long-term disease is that since we have become quite good at controlling diseases that kill us quickly, most of us will acquire lingering diseases with which we live for many years, especially since the process starts so early with attempts to control risk factors or pre-disease states. We may take expensive medications for prolonged periods and occasionally suffer more radical interventions that are even more expensive, while those of us who are really unlucky will live for some time in states of infirmity or as the French say *dépendance* where we cannot care for ourselves. Simultaneously, governments have taken on more and more financial and decision-making responsibility for our health and illnesses. Policymakers, in this telling, are simply seeking to adapt healthcare to this reality while trying desperately to keep a lid on rising costs and to satisfy increasingly vocal and demanding patient groups. This process is not new. Sixty years ago, the buzzwords in the United States were multiphasic screening, rehabilitation, coordination, and health promotion; they offered hope and some vision of how institutions might change. Now in our hyperconnected world it has taken only a few short years for chronic-care management, self-management, health education, and evidence-based illness protocols to spread across the developed world as both rhetoric and policy. And all this has occurred with little evidence that measures implementing such principles make a real difference.[46]

There is much that is new in this twenty-first-century concern with chronic disease. On both national and international levels, there now exist large health policy sectors where professionals interact, read the same literature, and share ideas and terminology. It is thus no accident that CCM appeared almost everywhere in the short period between 2001 and 2004. Patient groups are far more visible, influential, and demanding than in the past. Traditional personal responsibility for staying healthy has now been extended to management of illness. And faith in the clinical judgment of properly certified doctors and tolerance of treatment variability has been replaced by "evidence-based" efforts to control and standardize medical practice through guidelines and protocols while at the same time miraculously satisfying individual patient preferences.[47] But several things have not changed. One is the tendency to appeal to chronic disease "plagues" and "epidemics" in order to force change in stubbornly rigid institutional structures. Another is the tension between breaking down problems and programs into small, manageable units and the need for large categorical frameworks that integrate and seem to provide a semblance of control over the multiple problems, groups, and institutions that make up the massive, unwieldy patchworks that we call healthcare "systems." And there is one other thing that has barely changed. Despite the widespread rhetoric about chronic disease management, it remains a predominantly administrative notion that bears little relationship to what healthcare professionals mostly do or to what governments mostly pay for. The relatively small amounts of money now being spent can be justified by limited evidence about efficacy, but it is not clear that financing will ever be even remotely sufficient. Given that the thrust of policy everywhere is cost containment, money would have to be redirected away from classical medical services. The number of hospitals and hospital beds, it is true, has everywhere declined dramatically in the past few decades, but that is because hospitals stays keep getting shorter, tasks have been offloaded to other institutions or to families, and many countries are willing to accept long waiting lists for hospital services. The trend will surely continue but there may well be serious limits on how much hospital medicine we are willing or able to do without in order to finance management programs. Chronic disease thus remains what it has always been: not just a problem or series of problems to be solved and not merely a slogan but an amorphous and open-ended notion that conveys a wide variety of changing and sometimes conflicting fears, aspirations, and hopes.

Notes

Abbreviations

AAP-HP	Archives de l'Assistance Publique–Hôpitaux de Paris
AJN	*American Journal of Nursing*
AJPH	*American Journal of Public Health* (and *The Nation's Health*)
Am J Epid	*American Journal of Epidemiology*
ANF	Archives Nationales, Fontainebleau
Ann AAPSS	*Annals of the American Academy of Political and Social Science*
BMJ	*British Medical Journal*
Bull AM	*Bulletin de l'Académie de Médecine*
Bull Hist Med	*Bulletin of the History of Medicine*
Bull NYAM	*Bulletin of the New York Academy of Medicine*
Bull WHO	*Bulletin of the World Health Organization*
Int J Epid	*International Journal of Epidemiology*
JAMA	*Journal of the American Medical Association*
J Chron Dis	*Journal of Chronic Diseases*
J Hist Med Allied Sci	*Journal of the History of Medicine and Allied Sciences*
J R Coll Phys	*Journal of the Royal College of Physicians, London*
J R Soc Med	*Journal of the Royal Society of Medicine*
Milbank Q	*Milbank Quarterly, Milbank Memorial Fund Quarterly*
NCBI	National Center for Biotechnology Information
N Eng J Med	*New England Journal of Medicine*
NLM	National Library of Medicine
NYT	*The New York Times*
PHR	*Public Health Reports*
Pop Sc M	*Popular Science Monthly*
Proc NCSW	*Proceedings of the National Conference of Social Work*
Soc Hist Med	*Social History of Medicine*
Sociol Health Illn	*Sociology of Health and Illness*
Soc Sci Med	*Social Science and Medicine*
Yale J Biol Med	*Yale Journal of Biology and Medicine*

Note: For Internet sources, access dates are provided only when no date of publication or revision can be determined from the source title.

Introduction

1. D. M. Fox, "Health policy and changing epidemiology in the United States: chronic disease in the twentieth century," *Transactions & Studies of the College of Physicians of Philadelphia* 10 (1988): 11–31; idem, "Policy and epidemiology: financing health services for the chronically ill and disabled, 1930–1990," *Milbank Q* 67 suppl. 2, pt. 2 (1989): 257–87; idem, *Power and illness: the failure and future of American health policy* (Berkeley, 1993).

2. J. Szabo, *Incurable and intolerable: chronic disease and slow death in nineteenth-century France* (New Brunswick, NJ, 2009); S. Ebersold, *L'invention du handicap: la normalisation de l'infirme* (Paris, 1997); D. A. Stone, *The disabled state* (Philadelphia, 1984).

3. Aretaeus, *The extant works of Aretaeus, the Cappadocian: on the causes and symptoms of chronic disease, Book 1*, 49, ed. and trans. Francis Adams (Boston, 1972 [1856]).

4. For a survey of views over time, see A. Rabagliati, "Some remarks on the classification and nomenclature of diseases," *BMJ* 2 (1881): 114.

5. A. F. Chomel, "Chroniques (affections)," in MM. N. P. Adelon et al., *Dictionnaire de médecine ou répertoire général des sciences médicales*, 2nd ed. (Paris, 1834), 7:568.

6. W. Cadogan, *A dissertation on the gout, and all chronic diseases jointly considered, as proceeding from the same causes, what those causes are, and a rational and natural method of cure proposed: addressed to all invalids* (London, 1771), 11–14.

7. J. King, *The causes, symptoms, diagnosis, pathology, and treatment of chronic diseases* (Cincinnati, 1867).

8. S. Hahnemann, *The chronic diseases, their peculiar nature and their homœopathic cure: translated from the second edition of 1835 by Professor Louis A. Tafel* (Philadelphia, 1904), 42.

9. E.g., M. Durand-Fardel, *De la pathogénie des maladies chroniques au point de vue de la médication thermale, mémoire lu à la Société Impériale de Médecine de Lyon* (Lyon, 1856); "The principal baths of Germany, considered with reference to their remedial efficacy in chronic disease," *Literary Gazette* (1840): 540–41; A. S. Myrtle, *Chronic diseases best fitted for treatment by the Harrogate Mineral Springs* (London, 1876).

10. Anonymous review of *The prevention and cure of many chronic diseases by movement*, in *Provincial Medical & Surgical Journal* (1851): 631–32.

11. Among the earliest was M. La Beaume, *Remarks on the history and philosophy but particularly on the medical efficacy of electricity in the cure of nervous and chronic disorders* (London, 1820).

12. T. H. Levere, "Dr Thomas Beddoes and the establishment of his pneumatic institution: a tale of three presidents," *Notes and Records of the Royal Society of London* 32 (1977): 46.

13. E.g., T. M. Madden, *The principal health-resorts of Europe and Africa for the treatment of chronic diseases* (Philadelphia, 1876); A. J. Pleasonton, *The influence of the blue ray of the sunlight and of the blue color of the sky, in developing animal and vegetable life; in arresting disease and in restoring health in acute and chronic disorders to human and domestic animals* (Philadelphia, 1877).

14. E.g., C.-L. Dumas, *Doctrine générale des maladies chroniques pour servir de fondement à la connaissance théorique et pratique de ces maladies* (Paris, 1812); J. Poilroux, *Mé-*

moire qui a remporté le prix au jugement de la Société de médecine pratique de Montpellier sur la question proposée en ces termes: quel est le caractère distinctif des maladies chroniques? (Montpellier, 1812); M. Durand-Fardel, *Traité pratique des maladies chroniques*, 2 vols. (Paris, 1868).

15. Szabo, *Incurable and intolerable*, 32–33. His opening chapter constitutes the best discussion of nineteenth-century chronicity and incurability that is currently available. On academic debates generally, see G. Weisz, *The medical mandarins: the French Academy of Medicine in the nineteenth and early twentieth centuries* (Oxford, 1995), ch. 5.

16. Cited in E. M. Gruenberg, "The failures of success," *Milbank Q* 77 (1977): 3.

17. G. N. Grob, *The deadly truth: a history of disease in America* (Cambridge, MA, 2002), 218–20.

18. "The truth about cancer," *AJPH* 15 (1925): 338–39.

19. "Bibliography Bulletin 88, New York State Census Records, 1790–1925," compiled by Marilyn Douglas and Melinda Yates, New York State Library, October 1981, p. 44 (p. 50 of PDF). http://purl.org/net/nysl/nysdocs/9643270.

20. "Doctors will not reply; the census disease queries to be ignored," *NYT*, May 28, 1890; "Disease in the census," *Washington Post*, May 24, 1890.

21. H. B. Lang, "Social statistics? Defective physical conditions in Massachusetts from the decennial census of 1905," *Massachusetts Labor Bulletin* 12 (1907): 176–81; W. H. Mahoney, "Benevolent hospitals in Metropolitan Boston," *Publications of the American Statistical Association* 13 (1913): 419–48.

22. F. Condrau and M. Worboys, "Second opinions: epidemics and infections in nineteenth-century Britain," *Soc Hist Med* 20 (2007): 147–58; idem, "Epidemics and infections in nineteenth-century Britain," *Soc Hist Med* 22 (2009): 165–71; G. Mooney, "Infectious diseases and epidemiologic transition in Victorian Britain? Definitely," *Soc Hist Med* 20 (2007): 595–606.

23. J. E. Goldthwait, "An anatomic and mechanistic conception of disease," *Boston Medical and Surgical Journal* 172 (1915): 881–98; idem, *Essentials of body mechanics in health and disease* (Philadelphia, 1952).

24. The Surgeon-General's Catalogue, Series 2, for the years 1896 to 1916 contains one thousand-plus listings with the word *chronic* in the title. (The search function has a limit of one thousand hits.) Only seventy-two works have *chronic disease* in the title, and these are usually associated with specific joints or organs. This source can be accessed at www.nlm.nih.gov/hmd/indexcat/ichome.html.

25. One of the few examples is the French Law of 1905 providing free medical care to the indigent "incurables, infirm and elderly." The attempt to reform the Poor Laws in Britain at about the same time was not specifically aimed at this segment of the population. On the deleterious effects of exclusion in France, see Szabo, *Incurable and intolerable*.

26. J. M. Charcot and B. Ball, *Leçons sur les maladies des vieillards et les maladies chroniques* (Paris, 1868).

27. E.g., Fox, *Power and illness*, 22–25.

28. I owe this insight to David Armstrong, who first brought it to my attention.

29. W. R. Arney and B. J. Bergen, "The anomaly, the chronic patient and the play of

medical power," *Sociol Health Illn* 5 (1983): 1–24; idem, *Medicine and the management of living: taming the last great beast* (Chicago, 1984).

30. D. Armstrong, "Use of the genealogical method in the exploration of chronic illness: a research note," *Soc Sci Med* 30 (1990): 1225–27; idem, "The rise of surveillance medicine," *Sociol Health Illn* 17 (1995): 393–404; idem, "Chronic illness: epidemiological or social explosion?" *Chronic Illness* 1 (2005): 26–27; discussion 28–29.

31. D. Armstrong, "Chronic illness: a revisionist account," *Sociol Health Illn* article first published online: August 5, 2013, DOI: 10.1111/1467-9566.12037.

32. E.g., P. Conrad, "The shifting engines of medicalization," *Journal of Health and Social Behavior* 46 (2005): 3–14.

33. R. A. Aronowitz, *Unnatural history: breast cancer and American society* (Cambridge, 2007); I. Löwy, *Preventive strikes: women, precancer, and prophylactic surgery* (Baltimore, 2009).

CHAPTER ONE: "National Vitality" and Physical Examination

1. On the relationship between eugenics and other Progressive movements, see M. Ladd-Taylor, "Saving babies and sterilizing mothers: eugenics and welfare politics in the interwar United States," *Social Politics: International Studies in Gender, State & Society* 4 (1997): 136–53; W. Kline, *Building a better race: gender, sexuality, and eugenics from the turn of the century to the baby boom* (Berkeley, 2001), 13–14. On the international character of the Progressive movement, see D. T. Rodgers, *Atlantic crossings: social politics in a progressive age* (Cambridge, MA, 1998).

2. I. Fisher and the National Conservation Commission, *A report on national vitality, its wastes and conservation* (Washington, DC, 1909); R. W. Dimand, "Comments on William D. Nordhaus's, 'Irving Fisher and the contribution of improved longevity to living standards,'" *American Journal of Economics and Sociology* 64 (2005): 394–95.

3. See I. Fisher, "Lengthening human life in retrospect and prospect," *AJPH* 17 (1927): 1–14, for a short statement of his views.

4. L. D. Hirschbein, "Masculinity, work, and the fountain of youth: Irving Fisher and the Life Extension Institute, 1914–31," *Canadian Bulletin for the History of Medicine/ Bulletin Canadien d'histoire de la Médecine* 16 (1999): 91–92.

5. W. H. Peters, "Report of the Committee on Medical Inspection of Schools," *AJPH* 6 (1916): 589–91.

6. E. A. Toon, "Managing the conduct of the individual life: public health education and American public health, 1910 to 1940" (PhD diss., University of Pennsylvania, 1998), ch. 4 and pp. 227, 229.

7. K. Buhler-Wilkerson, "Care of the chronically ill at home: an unresolved dilemma in health policy for the United States," *Milbank Q* 85 (2007): 614–16.

8. Ibid., 617; S. C. Nelson, "Study of chronic cases," *Public Health Nurse* 21 (1929): 577–78.

9. Fisher, "Lengthening human life," 8.

10. Hirschbein, "Masculinity, work, and the fountain of youth," 96.

11. "Medicine: life extension," *Time Magazine*, April, 22, 1935. www.time.com/time/magazine/article/0,9171,762253,00.html.

12. L. K. Frankel, "Conservation of life by life insurance companies," *Ann AAPSS* 70 (1917): 77–78.

13. E. L. Fisk, "Some results of periodic health examinations," *Pop Sc M* 86 (1915): 324–30.

14. These include E. Sydenstricker and R. H. Britten, "General results of a statistical study of medical examinations by the Life Extension Institute of 100,924 white male life insurance policy holders since 1921," *American Journal of Hygiene* 11 (1930): 73–135; R. H. Britten, "Sex differences in the physical impairments of adult life: a comparison of rates among men and women, based on 112,618 medical examinations by the Life Extension Institute," *American Journal of Hygiene* 13 (1931): 741–70.

15. E. Marshall, "We waste life recklessly in this country," *NYT*, May 12, 1912.

16. E. E. Rittenhouse, "Preventable waste of life," *JAMA* 54 (1910): 294–95.

17. Ibid.; "Mortality record shows big increase," *NYT*, Feb. 3, 1915.

18. I. Fisher and E. L Fisk, *How to live: rules for healthful living, based on modern science*, 9th ed. (New York, 1916), 4–5.

19. Ibid., 17; Fisk, "Some results of periodic health examinations," 324.

20. H. M. Biggs, "Preventive medicine: its achievements, scope and possibilities. Oration on state medicine at the fifty-fifth annual session of the American Medical Association at Atlantic City, June 7–10, 1904," *JAMA* 42 (1904): 1550–55: quotes on 1551 and 1555 respectively.

21. Cited in D. Rosner, "Beyond Typhoid Mary: the origins of public health at Columbia and in the city" *Columbia Magazine* (2004). www.columbia.edu/cu/alumni/Magazine/Spring2004/publichealth.html; C.-E.A. Winslow, *Life of Hermann Biggs, MD, D. Sc., Ll. D., physician and statesman of public health* (Philadelphia, 1929), 241.

22. "30,000 a year slain by deadly economy; E.E. Rittenhouse brands city's neglect of Health Department as communal slaughter," *NYT*, March 13, 1913.

23. H. M. Biggs, "Practical objectives in health work during the next twenty years," *Proc NCSW* 50 (1923): 532.

24. Letter from S. S. Goldwater to C. L. Dana, Feb. 13, 1914, Papers of the Committee for Public Health, folder "1914–1933," NYAM archives.

25. "Subcommittee responds to Committee on Public Health, Hospitals, and Budget on the 'so called non-preventable diseases,'" report written by Nathan E. Brill and Floyd W. Crandall, April 6, 1914, NYAM archives.

26. Ibid.

27. Letter from Goldwater to the Committee, April 20, 1914, NYAM archives; J. Duffy, *A history of public health in New York City: 1866–1966* (New York, 1974), 276.

28. "Hygiene tests for all," *NYT*, May 3, 1914; "S.S. Goldwater wants every New Yorker physically examined yearly," *NYT*, May 10, 1914.

29. "Woods gives police a health booklet," *NYT*, Oct. 24, 1914.

30. C. Bolduan, "Haven Emerson, the public health statesman," *AJPH* 40 (1950): 2.

31. E. E. Rittenhouse, "Increasing organic disease—the new public health problem," *AJPH* 5 (1915): 1130–36.

32. Fisk, "Some results"; P.K.J. Han, "Historical changes in the objectives of the periodic health examination," *Annals of Internal Medicine* 127 (1997): 911.

33. Letter from Lewinski-Corwin to Miller, July 23, 1914, NYAM archives. Also see letters of July 23, 1914 (Brill to Miller), and July 25, 1914 (Lewinski-Corwin to Miller).

34. "Great need of the conservation of middle age; a really effective battle for probable health will begin when people realize the great waste of life among those who are past forty," *NYT*, Dec. 19, 1915.

35. Ibid. Much of this is also summarized in Lewinski-Corwin's letter to Miller, July 23, 1914, NYAM archives.

36. E. E. Rittenhouse, *America's pressing mortality problem: extraordinary increase in the death rate from organic disease of the heart and other hard worked organs, as indicated by the mortality records: urgent need of individual and government action* (New York, 1915).

37. Rittenhouse, "Increasing organic disease," 1134.

38. Ibid., 1134–35.

39. E. E. Rittenhouse, "Upbuilding national vitality—the need for a scientific investigation," *Science* 43 (1916): 222–23.

40. W. F. Willcox, "Fewer births and deaths: what do they mean?" *Journal of Heredity* 7 (1916): 119–27.

41. H. Emerson, "Reliability of statements of cause of death from the clinical and pathological viewpoints," *AJPH* 6 (1916): 685.

42. H. Emerson, "Maintenance of health in adults," *AJPH* 15 (1925): 705–6.

43. E. F. Bashford, "Fresh alarms on the increase of cancer," *Lancet* 183 (1914): 379–82. For responses, see ibid., 1079–82 and 1150–52.

44. L. I. Dublin, "Possibilities of reducing mortality at the higher age groups," *AJPH* 3 (1913): 1262–71.

45. Ibid., 1271.

46. L. I. Dublin, "Factors in American mortality," *American Economic Review* 6 (1916): 523–48; idem, "The trend of American vitality," *Pop Sc M* 86 (1915): 319.

47. C. B. Davenport, "The racial element in national vitality," *Pop Sc M* 86 (1915): 331–33.

48. L. I. Dublin, "The increasing mortality after age forty-five—some causes and explanations," *Publications of the American Statistical Association* 15 (1917): 512.

49. Ibid., 517, 522.

50. Ibid., 522.

51. L. I. Dublin and R. J. Vane, "Causes of death by occupation: occupational mortality experience of the Metropolitan Life Insurance Company, Industrial Department, 1922–1924," *U.S. Bureau of Labor Statistics Bulletin*, Industrial Accidents and Hygiene Series no. 507 (Washington, DC, 1930), 13.

52. "I told him he could add ten years to his life—and he laughed!" *NYT*, April 27, 1919.

53. L. K. Frankel, "The role of the life insurance company in health conservation programs," *Ann AAPSS* 130 (1927): 1–8.

54. E. L. Fisk, "The medical aspect of the changing status of the causes of sickness and death," *Proc NCSW* 55 (1928): 154–55; idem, "Possible extension of the human life cycle," *Ann AAPSS* 145 (1929): 153–201.

55. L. I. Dublin and A. J. Lotka, *Twenty-five years of health progress; a study of the mortal-*

ity experience among the industrial policyholders of the Metropolitan Life Insurance Company 1911 to 1935 (New York, 1937), 206.

56. Han, "Historical changes," 915; L. K. Frankel, "Presidential address of Lee K. Frankel, Ph.D.: presented to the Annual Meeting of the American Public Health Association, New Orleans, La., October 27th, 1919," *AJPH* 9 (1919): 811–18; The Boston Dispensary Records, 1871–1955 Charities Collection—CC 27, Simmons College archives, online description, http://my.simmons.edu/library/collections/college_archives/charities/char_coll _027.pdf. Accessed Sept. 13, 2012.

57. J. M. Dodson, "The American Medical Association and periodic health examinations," *AJPH* 15 (1925): 599–601.

58. Ibid., 600.

59. E. B. Edie, "Health examinations past and present and their promotion in Pennsylvania," *AJPH* 15 (1925): 605.

60. Ibid., 603.

61. A. N. Thomson, "The place of health examinations in the practice of medicine," *AJPH* 15 (1925): 608.

62. C.-E.A. Winslow, "Community defense of national vitality," *Pop Sc M* 86 (1915): 319–24; G. E. Vincent, "Teamplay in public health," *AJPH* 9 (1919): 14–21.

63. Frankel, "Presidential address, American Public Health Association," 811–18.

64. "Association news: A nation-wide health examination campaign," *AJPH* 13 (1923): 586.

65. J. A. Tobey, "The promotion of periodic health examinations: the Health Examination Campaign of the National Health Council," *Proc NCSW* 51 (1924): 225.

66. Ibid., 222.

67. F. A. Faught, "Health examinations: the relation of the nurse to periodic health examinations and life extension," *AJN* 27 (1927): 427–30.

68. G. Bates, "Preventive trend in the practice of medicine," *AJPH* 21 (1931): 249–52.

69. "To tell New Yorkers how to prolong lives," *NYT*, Aug. 5, 1929.

70. "Sees health tests as aid to doctors," *NYT*, Nov. 19, 1929.

71. "Sees step to save the family doctor," *NYT*, Dec. 29, 1929.

72. "Health examinations increase 25% here: rapid gain in tests reported by doctors following drive by medical society," *NYT*, April 15, 1930; "Plan 5-year drive for health tests; county medical groups here aim to create new course in the profession's schools," *NYT*, Aug. 17, 1930.

73. C.-E.A. Winslow, "Preventive medicine and health promotion—ideals or realities," *Yale J Biol Med* 14 (1942): 444.

74. Ibid., 447.

CHAPTER TWO: Expanding Public Health

1. C.-E.A. Winslow, "The untilled fields of public health," *Science* 51 (1920): 28.

2. E. Fee, *Disease and discovery: a history of the Johns Hopkins School of Hygiene and Public Health, 1916–1939* (Baltimore, 1987).

3. T. Parran, "Public responsibility for public and personal health: the Biggs Health Center Plan of 1920 in retrospect," *Bull NYAM* 11 (1935): 533–48; C.-E.A. Winslow, *Life*

of Hermann Biggs, MD, D. Sc., Ll. D., physician and statesman of public health (Philadelphia, 1929).

4. Parran, "Public responsibility," 535.

5. M. Terris, "Hermann Biggs' contribution to the modern concept of the health center," *Bull Hist Med* 20 (1946): 387–413.

6. Quoted in Terris, "Hermann Biggs' contribution," 396.

7. Ibid., appendix A, 402–13.

8. H. M. Biggs, "Practical objectives in health work during the next twenty years," *Proc NCSW* 50 (1923): 530–36.

9. W. T. Sedgwick, "American achievements and failures in public health work," *AJPH* 5 (1915): 1107.

10. F. P. Gay, "Whose business is the public health?" *Science* 54 (1921): 160.

11. E. C. Howe, "Organized hygiene: methods and results in instruction in personal hygiene—the prospect of improving the expectation of life in middle age," *AJPH* 6 (1916): 1039–48; L. K. Frankel, "Presidential address of Lee K. Frankel, Ph.D.: presented to the annual meeting of the American Public Health Association, New Orleans, October 27th, 1919," *AJPH* 9 (1919): 811–18.

12. F. L. Hoffman, "A plan for a more effective federal and state health administration," *AJPH* 9 (1919): 161–69. On Hoffman's controversial political views, see M. J. Wolff, "The myth of the actuary: life insurance and Frederick L. Hoffman's race traits and tendencies of the American Negro," *PHR* 121 (2006): 84–91; B. Hoffman. "Scientific racism, insurance, and opposition to the welfare state: Frederick L. Hoffman's transatlantic journey," *The Journal of the Gilded Age and Progressive Era* 2 (2003): 150–90.

13. A. Arkin, "The unsolved problems of preventive medicine," *AJPH* 11 (1921): 901.

14. "Association news—resolutions," *AJPH* 12 (1922): 1045–46.

15. I. V. Hiscock, "Charles-Edward Amory Winslow February 4, 1877–January 8, 1957," *Journal of Bacteriology* 73 (1957): 295–96; M. Terris, "C.-E.A. Winslow: scientist, activist, and theoretician of the American Public Health Movement throughout the first half of the twentieth century," *Journal of Public Health Policy* 19 (1998): 135–46: C.-E.A. Winslow, *Healthy living, the body and how to keep it well* (New York, 1917). The catalogue of the Library of Congress lists ten editions of this latter book, the last one published in 1935.

16. Winslow, *Healthy living*, 30.

17. Ibid., 32.

18. C.-E.A. Winslow, *Evolution and significance of the modern public health campaign* (New Haven, CT, 1923), 53.

19. Ibid., 54, 55.

20. Ibid., 58.

21. Ibid., 61.

22. C.-E. A. Winslow, "Public health at the crossroads," *AJPH* 16 (1926): 1077–78.

23. Ibid., 1078–79.

24. Ibid., 1079–80.

25. Ibid., 1080.

26. Ibid., 1083–84.

27. Cited in Winslow, *Life of Hermann Biggs*, 230.

28. G. E. Vincent, "Public welfare and public health," *Ann AAPSS* 105 (1923): 36, 41.

29. H. S. Cumming, "Developing a community health program," *Modern Hospital* 26 (1926): 4.

30. H. S. Cumming, "Public health and the public health service," *California and Western Medicine* 1 (1927): 36.

31. A. J. Viseltear, "Emergence of the Medical Care Section of the American Public Health Association, 1926–1948," *AJPH* 63 (1973): 986–1007.

32. "What is public health? A symposium," *AJPH* 18 (1928): 1019.

33. E. R. Kelley, "Cancer and the health administrator," *AJPH* 14 (1924): 561–62.

34. H. Emerson, "Maintenance of health in adults," *AJPH* 15 (1925): 705–6.

35. Ibid., 707.

36. Ibid.

37. H. Williams, "Current preoccupations of health officers," *AJPH* 23 (1933): 323–26.

38. C.-E. A. Winslow, "Steps in planning a health education and publicity program," *AJPH* 19 (1929): 650.

39. R. G. Leland, "The hospital as a public health agency in preventive medicine," *Modern Hospital* 26 (1926): 325–28.

40. W. M. Dickie, "Health examinations and the health officer," *AJPH* 15 (1925): 853–54.

41. Letter from I. Falk to J. Kingsbury and E. Sydenstricker, October 24, 1933, I. S. Falk Papers, MS 1039, box 38 #119, Yale University Library archives.

42. N. C. Erdey, "Armor of patience: The National Cancer Institute and the development of medical research policy in the United States, 1937–1971" (PhD diss., Case Western Reserve University, 1995), 11.

43. J. T. Patterson, *The dread disease: cancer and modern American culture* (Cambridge, MA, 1987).

44. First annual report of the executive secretary, Boston, 1922, Simmons College archives, http://my.simmons.edu/libraries/collections/college_archives/charities/char_coll_025.pdf.

45. Boston Council of Social Agencies, H. Emerson, and A. F. Hamburger, *Report on chronic disease in Boston, Mass* (Boston, 1927).

46. B. G. Rosenkrantz, *Public health and the state; changing views in Massachusetts, 1842–1936* (Cambridge, MA, 1972), 166–67.

47. Ibid., 167–68.

48. E. R. Kelley, L. I. Dublin, and M. P. Ravenel, "The control of cancer: report of the Committee on Control of Cancer, presented to the Public Health Administration Section of the American Public Health Association at the Fifty-third Annual Meeting at Detroit, Michigan, October 22, 1925," *AJPH* 15 (1925): 297–98.

49. G. A. Soper, G. H. Bigelow, and H. F. Vaughan, "What official public health agencies should do about cancer," *AJPH* 17 (1927): 1135–41.

50. Rosenkrantz, *Public health and the state*, 167.

51. H. L. Lombard, "Chronic disease problem in Massachusetts," *Proc NCSW* 57 (1930): 146.

52. G. H. Bigelow, "The Massachusetts Cancer Program," *Proc NCSW* 57 (1930): 155, 162.

53. R. B. Osgood, "The governor's welfare program and certain aspects of chronic disease," *N Eng J Med* 202 (1930): 527–28.

54. G. H. Bigelow and H. L. Lombard, *Cancer and other chronic diseases in Massachusetts* (Boston, 1933), 1.

55. Ibid., 5–6, 29.

56. Ibid., vii–viii.

57. Ibid., 4.

58. Ibid., 21–22.

59. Ibid., 75–76.

60. H. Emerson and A. C. Phillips, *Report of a study of the organized care of the sick, and of health agencies of Boston, for the Citizen's Committee of the Emergency Campaign of 1934, the Boston Council of Social Agencies, and the Hospital Superintendent's Club, September–October, 1934* (Boston, 1934).

61. V. A. Harden, *Inventing the NIH: federal biomedical research policy 1887–1937* (Baltimore, 1986), 160.

62. Harden, *Inventing the NIH*, 174.

63. D. Swain, "The rise of a research empire: NIH, 1930–1950," *Science* 138 (1962): 1234; Harden, *Inventing the NIH*, 173–74.

64. J. F. Jekel. "Health departments in the US 1920–1988: statements of mission with special reference to the role of C.-E.A. Winslow," *Yale J Biol Med* 64 (1991): 467–79.

65. T. Parran, "Cancer and the public health," *Science* 90 (1939): 428.

CHAPTER THREE: Almshouses, Hospitals, and the Sick Poor

1. B. Linker, *War's waste: rehabilitation in World War I America* (Chicago, 2011).

2. R. W. Bruère, "Some recent experiments in human conservation," *Harper's Magazine* (March 1911): 515–23.

3. C. L. King, "Foreword," *Ann AAPSS* 105 (1923): vi.

4. H. W. Odum, "Newer ideals of public welfare," *Ann AAPSS* 105 (1923): 2.

5. Ibid., 4, 6.

6. R. B. Rankin, "Department of Public Welfare in the City of New York," *Ann AAPSS* 105 (1923): 153.

7. G. E. Vincent, "Public welfare and public health," *Ann AAPSS* 105 (1923): 41.

8. "Social work and public health," *AJPH* 12 (1922): 702–3; emphasis in original.

9. E. K. Abel, "'In the last stages of irremediable disease': American hospitals and dying patients before World War II," *Bull Hist Med* 85 (2011): 29–56.

10. Ibid., 46.

11. Cleveland Hospital Council, *Cleveland hospital and health survey, hospitals and dispensaries, part ten* (Cleveland, 1920), 951.

12. B. B. Perkins, "Designing HIGH-COST medicine hospital surveys, health planning, and the paradox of progressive reform," *AJPH* 100 (2010): 223–33.

13. A. C. Bachmeyer, "What the survey shows about hospital conditions in Cincinnati," *Modern Hospital* 25 (1925): 338–39.

14. H. Emerson, "Organization of social forces: health and hospital surveys: Louisville's case history 1924," *Proc NCSW* 52 (1925): 461.

15. E.g., H. Emerson and A. C. Phillips, *Health and hospital survey of Bethlehem, Pennsylvania, 1925* (Bethlehem, 1926); H. Emerson, A. C. Phillips, and Community Chest Louisville Ky., *Health and Hospital Survey Committee, hospitals and health agencies of Louisville, 1924* (Louisville, 1925).

16. E. H. Corwin-Lewinski, *The hospital situation in Greater New York: report of a survey of hospitals in New York City* (New York, 1924).

17. Letter and memorandum of Sept. 26, 1924, from Ernst Boas to Charles Dana, Council on Public Health Papers, NYAM archives.

18. E. C. Potter, "Care of the chronically ill and the dependent," *Nation's Health* 7 (1925): 659–61.

19. N. R. Deardorff, "Social study by councils of social agencies and community chests," *Social Service Review* 11 (1937): 167–94.

20. H. Emerson and A. F. Hamburger, *Report on chronic disease in Boston, Mass* (Boston, 1927); also in *N Eng J Med* 199 (1928): 556–632.

21. H. W. Green, "An analysis and classification of Cleveland chronics," *Nation's Health* 9 (1927): 17–21.

22. N. R. Deardorff, "The social studies of the Welfare Council of New York City," *Journal of Educational Sociology* 4 (1931): 585–94.

23. On Jarrett, see J. M. Gabriel, "Mass-producing the individual: Mary C. Jarrett, Elmer E. Southard, and the industrial origins of psychiatric social work," *Bull Hist Med* 79 (2005): 430–58.

24. M. M. Davis and M. C. Jarrett, *A health inventory of New York City: a study of the volume and distribution of health service in the five boroughs* (New York, 1929); M. C. Jarrett, *The care of the chronic sick in private homes for the aged in and near New York City* (New York, 1931); idem, *Chronic illness in New York City*, 2 vols. (New York, 1933).

25. New Jersey, Dept. of Institutions & Agencies, *Report on chronic disease in New Jersey* (Trenton, NJ, 1932).

26. E. P. Boas and N. Michelsohn, *The challenge of chronic diseases* (New York, 1929).

27. Emerson and Hamburger, *Report on chronic disease in Boston*, 560, 622.

28. Potter, "Care of the chronically ill," 659.

29. Green, "An analysis of Cleveland chronics," 18.

30. New Jersey, *Report on chronic disease*, 522.

31. Jarrett, *Chronic illness in New York City*, 1:12, 18–20; G. N. Grob, *The deadly truth: a history of disease in America* (Cambridge, MA, 2002), 228–29.

32. T. S. Kerson, "Almshouse to municipal hospital: the Baltimore experience," *Bull Hist Med* 55 (1981): 203–20; Michael B. Katz, "Poorhouses and the origins of the public old age home," *Milbank Q* 62 (1984): 110–40. Also see works cited in notes 33 and 34.

33. C. E. Rosenberg, "From almshouse to hospital: the shaping of Philadelphia General Hospital," *Milbank Q* 60 (1982): 114.

34. M. Holstein and T. R. Cole, "The evolution of long-term care in America," in *The future of long-term care: social and policy issues*, ed. R. H. Binstock, L. E. Cluff, and O. Von Mering (Baltimore, 1996), 28.

35. Cited in ibid.

36. E. Schell, "The origins of geriatric nursing: the chronically ill elderly in almshouses and nursing homes, 1900–1950," *Nursing History Revue* 1 (1993): 206–8.

37. Katz, "Poorhouses," suggests that this evolution was deliberate and was at the origins of public old age homes. Holstein and Cole, "Evolution of long-term care," in contrast, argue that hatred of almshouses led to resistance to public provision of nursing home care, leading to the proprietary nursing home industry.

38. "Finds almshouses filled by disease," *NYT*, Jan. 2, 1928.

39. Boas and Michelsohn, *Challenge of chronic diseases*, 47–48, 51, 68.

40. Rosenberg, "From almshouse to hospital," 146.

41. Emerson, "Organization of social forces," 478.

42. Ibid. See also R. Sartwell, "How can we best care for the chronic?" *Modern Hospital* 27 (1926): 67–70.

43. Rosenberg, "From almshouse to hospital," 149–50.

44. Kerson, "Almshouse to municipal hospital," 217; Rosenberg, "From almshouse to hospital," 115.

45. Emerson, "Organization of social forces," 478.

46. W. H. Mahoney, "Benevolent hospitals in Metropolitan Boston," *Publications of the American Statistical Association* 13 (1913): 419–48; Rosenberg, "From almshouse to hospital," 128.

47. Harry F. Dowling, *City hospitals: the undercare of the underprivileged* (Cambridge, MA, 1982), 153. See also 154–55.

48. E. H. Lewinski-Corwin, "Community responsibility of hospitals," *Transactions of the American Hospital Association* 27 (1925): 510.

49. "Thousands visit Montefiore Home," *NYT*, May 18, 1914.

50. "In a model hospital," *NYT*, Aug. 27, 1922; Diana Rice, "Montefiore Hospital has fiftieth birthday," *NYT*, Dec. 2, 1934.

51. "The most fatal maladies," *NYT*, April 16, 1922.

52. "Montefiore Hospital marking 50th year," *NYT*, Dec. 3, 1934.

53. "Give $200,000 for private hospital," *NYT*, June 26, 1912.

54. F. H. McCrudden, "Scientific research in chronic medicine from the physiological point of view: the work of the Robert B. Brigham Hospital," *Boston Medical and Surgical Journal* 175 (1916): 129. The two quotes that follow in this paragraph are also on this page.

55. Ibid., 130.

56. A. C. Jensen, "Where chronic patients are receiving special consideration," *Modern Hospital* 26 (1926): 307.

57. A. C. Jensen, "What Fairmont Hospital is doing for the chronic and the convalescent," *Modern Hospital* 28 (1927): 61–67; C. W. Munger, "The modern hospital reading course: lesson XXII—care for chronics, convalescents and the preventorium child," *Modern Hospital* 31 (1928): 97–101.

58. H. Smith, "Extending existing hospital facilities in community or private hospitals to chronic and incurable patients," *Transactions of the American Hospital Association* 34 (1932): 634–35.

59. Boas and Michelsohn, *Challenge of chronic diseases*, 3.

60. Ibid., 7.

61. "Warns against word 'incurable,'" *NYT*, Oct. 13, 1927.

62. Boas and Michelsohn, *Challenge of chronic diseases*, 10, 11.

63. Ibid., 14–16.

64. On the general lack of interest in homecare during this period, see A. E. Benjamin, "An historical perspective on home care policy," *Milbank Q* 71 (1993): 129–66.

65. Ernst P. Boas, *The unseen plague: chronic disease* (New York, 1940), 47. See the review in *JAMA* 115 (1950): 2021.

66. Boas and Michelsohn, *Challenge of chronic diseases*, 88–89.

67. M. Constantine, "The nursing of chronic diseases—what it demands and what it offers," *Modern Hospital* 27 (1927): 128–32.

68. "Cancer, beware!" *Time Magazine*, June 9, 1924; I. Levin, "The cancer problem and the nurse," *AJN* 27 (1927): 87.

69. Letter from E. Boas to C. Dana, Sept. 26, 1924, NYAM archives.

70. J. Duffy, *A history of public health in New York City: 1866–1966* (New York, 1974), 300–302.

71. Jarrett, *Chronic illness in New York City*, 1:7.

72. Ibid., 1:60–63.

73. Letter from E. Boas to J. A. Miller, Dec. 28, 1933, NYAM archives.

74. "New board to aid chronically sick," *NYT*, March 25, 1934.

75. "The care of the chronically ill by the City of New York," memorandum prepared by Committee on Chronic Illness on Dec. 18, 1933, NYAM archives.

76. Bluestone's untitled report sent to the chairman of the NYAM Public Health Relations Committee, addressed to J. A. Miller, March 20, 1934, NYAM archives. See especially pp. 7, 11. See also E. M. Bluestone, "Integrating acute and chronic in a combined hospital-home pattern," *Lancet* 258 (1951): 78–80.

77. "Hospital is urged for chronic cases," *NYT*, May 21, 1934.

78. On this last point, see M. C. Jarrett, "The plight of the chronic patient," *Modern Hospital* 45 (1935): 71.

79. S. S. Goldwater, "Crusading for the chronically sick," *Modern Hospital* 44 (1935): 65–67.

80. "Mayor lays stone of new hospital," *NYT*, Oct. 6, 1937; "1,500 at preview of New Hospital," *NYT*, Feb. 5, 1939. For the mural discussed below, see www.americanabstract artists.org/history/wpamurals/hospital.html. Accessed Nov. 28, 2012.

81. C. A. Ford, "A community charge—the chronic," *Modern Hospital* 48 (1937): 83.

82. Boas, *The unseen plague*, 85; M. C. Jarrett, "Chronically ill and aged: standards for care for these patients," *Modern Hospital* 51 (1938): 45.

83. "Note," *Modern Hospital* 44 (1935): 106; "Private aid asked for city hospitals," *NYT*, April 11, 1935. There are several letters regarding funding for this initiative in Louis I. Dublin Papers, MS C 316 box 7, "Chronic Diseases, 1922–1940," NLM archives.

84. Memo of March 28, 1937, and letter from Dublin to Goldwater, March 30, 1937, both in Dublin Papers, box 7, NLM archives.

85. "City research post goes to Dr. Seegal," *NYT*, April 24, 1936; "Disease study aided by Rockefeller Fund; foundation gives city $66,000 to help research project on Welfare

Island," *NYT*, Feb. 1, 1938; "Research on chronic diseases at Welfare Island," *Science* 87 (1936): 131–32.

86. "Chronic ills held our worst plague," *NYT*, July 30, 1939.

87. "Battle spurred on chronic ills; diorama, set up on Grand Central Station balcony," *NYT*, July 25, 1939.

88. "Chronic ills held our worst plague."

89. American Public Welfare Association, board minutes June 12, 1930, in "History of the APWA," University of Minnesota archives website, http://special.lib.umn.edu/findaid/xml/sw0054.xml.

90. N. L. Williams, *Public welfare agencies and hospitals: a study in relationships* (Chicago, 1937).

91. American Hospital Association and American Public Welfare Association, *Institutional care of the chronically ill: a report of the Joint Committee on Hospital Care of American Hospital Association and American Public Welfare Association / Michael M. Davis, Chairman* (Chicago, 1940).

CHAPTER FOUR: New Deal Politics and the National Health Survey

1. "Laskers give $1,000,000 to U of C," *Chicago Daily Tribune*, Jan. 9, 1928; E. F. Barnard, "Science attacks slow ills of old age," *NYT*, Feb. 12, 1928.

2. E. P. Boas, *The unseen plague: chronic disease* (New York, 1940), 63.

3. J. T. Patterson, *The dread disease: cancer and modern American culture* (Cambridge, MA, 1987), 94; C. Hayter, "Cancer: the worst scourge of civilized mankind," *Canadian Bulletin of Medical History* 20 (2003): 261.

4. Patterson, *The dread disease*, 117; "Millions for cancer," *Time Magazine*, July 5, 1937, 37.

5. V. A. Harden, *Inventing the NIH: federal biomedical research policy 1887–1937* (Baltimore, 1986), 274; D. Swain, "The rise of a research empire: NIH, 1930–1950," *Science* (New Series) 138 (1962): 1234.

6. C. Sadler, "'Conquer Cancer' adopted as battle cry of the Public Health Service," *Washington Post*, Aug. 8, 1937; D. Cantor, "Radium and the origins of the National Cancer Institute," in *Biomedicine in the twentieth century: practices, policies, and politics*, ed. C. Hannaway (Amsterdam, 2008), 95–146.

7. Memo from Leonard Scheele to C. E. Waller, June 16, 1939, RG 443, box 192, National Archives. This was followed by a model letter to be sent to all public health regional consultants.

8. Some of its complex organizational features are described in P. J. Funigiello, *Chronic politics: health care security from FDR to George W. Bush* (Lawrence, KS, 2005), 24–29.

9. H. S. Cumming, "Chronic disease as a public health problem," *Milbank Q* 14 (1936): 127; emphasis mine.

10. J. Eyler, "Health statistics in historical perspective," in *Health statistics: shaping policy and practice to improve the population's health*, ed. D. J. Friedman, E. L. Hunter, and R. G. Parrish (Oxford, 2005), 42.

11. I. S. Falk, "The Committee on the Costs of Medical Care—25 years of progress,"

AJPH 48 (1958): 979–82; M. I. Roemer, "I. S. Falk, the Committee on the Costs of Medical Care, and the drive for national health insurance," *AJPH* 75 (1985): 841–48. Falk's more personal reminiscences are in L. E. Weeks and H. J. Berman, *Shapers of American health care policy: an oral history* (Ann Arbor, 1985), 12–21.

12. This and other material that follows is based on documents in the Isidore S. Falk Papers, Yale University, Sterling Memorial Library, Manuscripts and Archives, MS 1039 of the Contemporary Medical Care and Health Policy Collection. Falk eventually published a book on this subject himself. I. S. Falk, *Security against sickness: a study of health insurance* (Garden City, NY, 1936).

13. "Social Security online: the Committee on Economic Security," www.ssa.gov/history/reports/ces/cesbasic.html. Accessed Dec. 1, 2009.

14. "Social Security online: members of the committee, advisory boards and staff," www.ssa.gov/history/reports/ces/ces6.html. Accessed Dec. 1, 2009.

15. "Committee on Economic Security—medical advisory board: minutes of meetings, part 2—Tuesday afternoon session, January 29, 1935," www.ssa.gov/history/reports/ces/ces7minutes2.html. Also see T. Parran, "Health security," *Milbank Q* 14 (1936): 113–24.

16. E. Sydenstricker, "Public health provisions of the Social Security Act," *Law and Contemporary Problems* 3 (1936): 264.

17. C. E. Waller, "The Social Security Act in its relation to public health," *AJPH* 25 (1935): 1186–94.

18. T. Parran, "Health security," *Milbank Q* 14 (1936): 113–24.

19. T. Parran, "Reporting progress: presidential address," *AJPH* 26 (1936): 1076; interview with Perrott by Peter Corner, 17, George St. John Perrott Papers, MS 370, box 1, folder 4, 1966, NLM archives.

20. Swain, "The rise of a research empire," 1233–34.

21. Waller, "The Social Security Act," 1192.

22. Letter from Sydenstricker to Falk, June 27, 1935, Falk Papers, 39 #135, Yale University archives. Perrott later expressed similar views. interview by Peter Corner, 17, Perrott Papers.

23. Waller, "The Social Security Act," 1192.

24. Ibid., 1193.

25. "Executive Order of the President creating the Committee," cited in C. E. Walker, "The National Health Program," *Journal of the National Medical Association* 30 (1938): 147–51.

26. Transcript, oral history interview conducted by Peter A. Corning, part 1 (copyright, Columbia University) 1965, Falk Papers, series 5, box 172; file 2634, pp. 196–97, Yale University archives.

27. Associated Press, "New Deal surveys life on 700 fronts," *NYT*, Jan. 12, 1936.

28. Letter of Aug. 22, 1935, Sydenstricker to Falk, Falk Papers, series 2, box 39, file 133, Yale University archives.

29. "A progress report on a plan of analysis for the survey of chronic disease in the United States," Falk Papers, series 2, box 39, file 139, Yale University archives.

30. Interview with Perrott by Peter Corner 1966, pp. 21–23.

31. Letter of May 28, 1935, from Falk to L. Thomson, Falk Papers, series 2, box 43, file 321, Yale University archives. A copy also exists in the National Archives: NIH, RG443, box 185.

32. Cumming, "Chronic disease as a public health problem," 128.

33. P. J. Funigiello, *Chronic politics: health care security from FDR to George W. Bush* (Lawrence, 2005), 24.

34. Memo from "Operating Council," March 6, 1937, Records of the National Institutes of Health (NIH), RG 443, box 185, National Archives. In the same box there is material, including a letter from Josephine Roche to Thomas Parran, June 26, 1937, discussing another conflict over the laying off of workers now that survey work was completed.

35. Perrott interview with Corning, 13; Parran memo to L. Thomson, July 7, 1937, NIH, RG 443, box 185, National Archives.

36. Numerous materials of this sort are in RG 443, boxes 1 and 2. Among those in box 2 is E. Snyder, "The press and the National Health Survey," *Health Officer* 2 (1938): 552–54.

37. N. Krieger, *Embodying inequality: epidemiologic perspectives* (New York, 2005), 39–42.

38. E. Sydenstricker, "A study of illness in a general population group: Hagerstown morbidity studies no. I: the method of study and general results," *PHR* 41 (1926): 2083–84.

39. E. Sydenstricker, *Health and environment* (New York, 1972 [1933]).

40. Ibid., 86–88.

41. E. Sydenstricker, "Economic status and the incidence of illness: Hagerstown morbidity studies no. X: gross and specific illness rates by age and cause among persons classified according to family economic status," *PHR* 44 (1929): 1827.

42. Ibid.

43. I. S. Falk, M. C. Klem, and N. Sinai, *The incidence of illness and the receipt and costs of medical care among representative families* (New York, 1976 [1933]).

44. Sydenstricker, *Health and environment*, 38.

45. S. D. Collins, "The incidence and causes of illness at specific ages," *Milbank Q* 13 (1935): 320–38.

46. G.S.J. Perrott and S. D. Collins, "Relation of sickness to income and income change in 10 surveyed communities: Health and Depression Studies No. 1: method of study and general results for each locality," *PHR* 50 (1935): 595–622. On Sydenstricker's exclusion of African Americans from his health studies, and on his early interest in the economic burden of illness, see H. M. Marks, "Epidemiologists explain pellagra: gender, race, and political economy in the work of Edgar Sydenstricker," *J Hist Med Allied Sci* 58 (2003): 34–45.

47. Perrott and Collins, "Relation of sickness to income," 595.

48. Ibid., 607.

49. Ibid., 607, 619.

50. Ibid., 622.

51. J. Roche, "Cost of Depression in health revealed," *NYT*, Sept. 15, 1935.

52. Ibid.

53. L. I. Dublin and A. J. Lotka, *Twenty-five years of health progress; a study of the mortal-

ity experience among the industrial policyholders of the Metropolitan Life Insurance Company 1911 to 1935 (New York, 1937), 184–85, 272–73, 283–84.

54. Sydenstricker, *Health and environment*; "Social Security online: Committee on Economic Security Report on Health Insurance: the unpublished 1935 report on health insurance & disability, March 7, 1935, www.socialsecurity.gov/history/reports/health.html.

55. G.S.J. Perrott and H. C. Griffin, "An inventory of the serious disabilities of the urban relief population," *Milbank Q* 14 (1936): 216.

56. Ibid., 218.

57. E. Sydenstricker, "Surgeon general's report on health and Depression study," *Milbank Q* 14 (1936): 207.

58. Ibid., 207–8.

59. G.S.J. Perrott, "The state of the nation's health," *Ann AAPSS* 188 (1936): 131–43.

60. E. Sydenstricker, *The challenge of facts: selected public health papers of Edgar Sydenstricker*. (New York, 1974), 169.

61. Sydenstricker, "Study of illness in a general population group," 2074; emphasis in original.

62. Ibid.

63. E. Sydenstricker, "The incidence of various diseases according to age: Hagerstown morbidity studies VIII," *PHR* 43 (1928): 1125.

64. Perrott, "The state of the nation's health," 138.

65. Ibid., 139–40.

66. Ibid., 140–41.

67. F. E. Linder, "National Health Survey," *Science* 127 (1958): 1277.

68. J. Roche, "Economic health and public health objectives," *AJPH* 25 (1935): 1183.

69. National Health Survey, "Preliminary reports, bulletin 6: the magnitude of the chronic disease problem in the United States," in *Sickness and medical care series* (Washington, DC, 1938 [revised 1939]), 2.

70. Ibid., 12.

71. *Time Magazine*, Jan. 31, 1938, www.time.com/time/magazine/article/0,9171,759043,00.html.

72. J. S. Whitney, *Death rates by occupation (based on data of the U.S. Census Bureau, 1930)*, New York, 1934, cited in G.S.J. Perrott and D. F. Holland, "Health as an element in Social Security," *Ann AAPSS* 202 (1939): 120.

73. Ibid., 121–22.

74. D. F. Holland and G.S.J. Perrott, "Health of the Negro," *Milbank Q* 16 (1938): 5–38, especially 34. Republished *Milbank Q* 83 (2005).

75. R. H. Britten, "Receipt of medical services in different urban population groups, the National Health Survey," *PHR* 55 (1940): 2199–2224.

76. Perrott and Holland, "Health as an element," 123.

77. Ibid.

78. Ibid., 128.

79. Ibid., 125.

80. G.S.J. Perrott and D. F. Holland, "Population trends and problems of public health," *Milbank Q* 18 (1940): 359–92.

81. Letter from Owen West to Hugh Cummings, Sept. 10, 1935, NIH, RG 443, box 2, National Archives.

82. Perrott interview with Corning I—IV, 30, 36, NLM archives.

83. "Fishbein urges Department of Health for U.S," *Washington Post*, Feb. 11, 1938.

84. Ibid.; and Craig Thompson, "Urges doctors in social medicine," *NYT*, June 15, 1938.

85. "Current comment," *JAMA* 116 (1941): 1912.

86. C. C. Lienau, "Selection, training and performance of the National Health Survey field staff," *American Journal of Epidemiology* 34 (1941): 110–32.

87. Anonymous, "Personnel factors in the National Health Survey," *JAMA* 119 (1942): 347.

88. C. C. Lienau, "Personnel factors in the National Health Survey," *JAMA* 119 (1942): 968.

89. H. Emerson, "The physician's part in organized medical care," *AJPH* 30 (1940): 15.

90. Memo from Philip S. Broughton to Thomas Parran, Jan. 12, 1938, RG 443, box 193, National Archives.

91. "Social Security online: The need for a national health report of the Technical Committee on Medical Care (Washington, 1938)," www.ssa.gov/history/reports/Inter departmental.html.

92. Ibid., 3–4, 15, 33.

93. J. Downes, "Findings of the study of chronic disease in the Eastern Health District of Baltimore," *Milbank Q* 22 (1944): 337–51; D. E. Lilienfeld, "Harold Fred Dorn and the first National Cancer Survey (1937–1939): the founding of modern cancer epidemiology," *AJPH* 98 (2008): 2150–58.

94. D. M. Fox, Review of Philip J. Funigiello, *Chronic politics: health care security from FDR to George W. Bush* (Lawrence, 2005), *American Historical Review* 112 (2007): 248–49. See also E. D. Berkowitz, *Rehabilitation: the federal government's response to disability, 1935–1954* (New York, 1980), 43.

95. American Hospital Association and American Public Welfare Association, *Institutional care of the chronically ill: a report of the Joint Committee on Hospital Care of American Hospital Association and American Public Welfare Association / Michael M. Davis, Chairman* (Chicago, 1940).

96. E. P. Boas and N. Michelsohn, *The challenge of chronic diseases* (New York, 1929); E. P. Boas, *The unseen plague: chronic disease* (New York, 1940).

CHAPTER FIVE: Mobilizing against Chronic Illness at Midcentury

1. See, e.g., A. Derickson, " 'Health for three-thirds of the nation': public health advocacy of universal access to medical care in the United States," *AJPH* 92 (2002): 187.

2. CCI, *The Commission on Chronic Illness; origin, structure, objectives, staffing and financing* (Chicago, 1949): 7; F. Jensen, H. G. Weiskotten, and M. A. Thomas, *Medical care of the discharged hospital patient* (New York, 1944).

3. United States, President's Commission on the Health Needs of the Nation, *Building America's health: a report to the president* (Washington, DC, 1953), 9.

4. J. A. Greene, *Prescribing by numbers: drugs and the definition of disease* (Baltimore, 2007); V. Quirke, "From evidence to market: Alfred Spinks' 1953 survey of fields for phar-

macological research, and the origins of ICI's cardiovascular programme," in *Medicine, the market and the mass media: producing health in the twentieth century*, ed. V. Berridge and K. Loughlin (London, 2005), 146–71.

5. A. Derickson, *Health security for all: dreams of universal health care in America* (Baltimore, 2005), 92–93, 99–100.

6. E.g., N. Tomes, "Merchants of health: medicine and consumer culture in the United States, 1900–1940," *Journal of American History* 88 (2001): 519–47.

7. R. Stevens, *In sickness and in wealth: American hospitals in the twentieth century (updated edition)* (Baltimore, 1999), 214.

8. American Hospital Association and American Public Welfare Association, *Institutional care of the chronically ill: a report of the Joint Committee on Hospital Care of American Hospital Association and American Public Welfare Association, Michael M. Davis, Chairman* (Chicago, 1940).

9. Letter from Howard Russell to Claude Pepper, March 28, 1945, Ernst Boas Papers, American Philosophical Society; "Senators demand more health aids," *NYT*, Feb. 26, 1945.

10. Several accounts place the AMA at the initial November meeting but this was clearly not the case. See the annual report for 1946 of the APWA, SW054, box 1, University of Minnesota archives.

11. "'Planning for the chronically ill': a joint statement of recommendations by the American Hospital Association, American Medical Association, American Public Health Association and American Public Welfare Association," *AJPH* 37 (1947): 1256–66. All quotes are on p. 1256. Other publications that printed the document were *JAMA* and *Public Welfare*.

12. Quotes in this paragraph, ibid., 1257.

13. Quotes in this paragraph, ibid., 1258.

14. Quotes in this paragraph, ibid., 1258–61.

15. Quotes in this paragraph, ibid., 1261–62.

16. Ibid., 1262–63.

17. Quotes in this paragraph, ibid., 1265–66.

18. E.g., *APWA Annual Report 1947*, SWO54, box 1, p. 3, University of Minnesota archives.

19. M. Terris, "National planning for the chronically ill," *Proc NCSW* 57 (1948): 205.

20. Ibid., 207.

21. Cited in D. W. Roberts, "The Commission on Chronic Illness," *PHR* 69 (1954): 295.

22. National Health Assembly, *America's health: a report to the nation* (New York, 1949).

23. CCI, *Chronic illness in the United States: vol. 1, prevention of chronic illness* (Cambridge, MA, 1957), 291–92; CCI, *The Commission on Chronic Illness*, 13.

24. M. L. Levin, H. Goldstein, and P. R. Gerhardt, "Cancer and tobacco smoking," *JAMA* 143 (1950): 336–38. The story was told in M. Terris, "Re 'Morton Levin (1904–1995): history in the making,'" *Am J Epid* 146 (1997): 365.

25. CCI, "Second annual meeting, May 8, 1950, Chicago," Leonard Mayo Papers, box 6, p. 7, University of Minnesota archives.

26. Roberts, "Commission on Chronic Illness," 295.

27. CCI, "Second annual meeting," 6–8.

28. Roberts, " Commission on Chronic Illness," 296; CCI, *Chronic illness in the United States: vol. 1*, front pages.

29. Boas to Levin, undated letter; Levin to Boas, April 6, 1951, Ernst Boas Papers, American Philosophical Society.

30. CCI, "Minutes of the Executive Committee, October 26, 1951," Ernst Boas Papers, American Philosophical Society.

31. CCI, *Chronic illness in the United States: vol. 1*, 298, 307–8.

32. CCI, "Minutes of Executive Committee Meeting, June 15, 1952," Ernst Boas Papers, American Philosophical Society.

33. CCI, *Proceedings of the first meeting* (Chicago, 1949), 3–4.

34. Ibid., 9.

35. Ibid., 14.

36. Ibid.

37. Ibid., 17–19.

38. Ibid., 294. On state medical societies, see "Activities of state medical societies in chronic illness planning," *JAMA* 144 (1950): 1262–63.

39. "The monthly newsletter," "Second meeting, CCI," Leonard Mayo Papers, box 6, pp. 39–40, University of Minnesota archives; "CCI, report on educational activities—1950," Ernst Boas Papers, American Philosophical Society.

40. This summary is based on CCI, *Chronic illness in the United States: vol. 1*, 295–304. Among the reports mentioned here are US Public Health Service and Commission on Chronic Illness, *A study of selected home care programs; a joint project of the Public Health Service and the Commission on Chronic Illness*, Public Health Monograph no. 35 (Washington, DC, 1955), 447; J. Solon, D. W. Roberts, D. E. Krueger, and A. M. Baney, *Nursing homes, their patients and their care: a study of nursing homes and similar long-term care facilities in 13 states*, Public Health Monograph no. 46 (Washington, DC, 1957); V. B. Turner and US Public Health Service, *Chronic illness: digest of selected reference* (Washington, DC, 1951); Public Health Service and Commission on Chronic Illness, *Chronic disease hospitals; reports from 12 selected institutions* (Washington, DC, 1954).

41. CCI, *Chronic illness in the United States: vol. 1*; CCI, *Chronic illness in the United States: vol. 2, care of the long-term patient* (Cambridge, MA, 1957).

42. United States, *Building America's health: a report to the president* (Washington, DC, 1953).

43. CCI, "Minutes of Executive Committee, September 11, 1952," Ernst Boas Papers, American Philosophical Society.

44. *Building America's health*, table of contents, xiv.

45. CCI, "Minutes of Executive Committee, September 11, 1952," Ernst Boas Papers, American Philosophical Society; emphasis mine.

46. CCI, *Chronic illness in the United States: vol. 1*, 21.

47. CCI, *Chronic illness in the United States: vol. 2*, 425.

48. CCI, *Chronic illness in the United States: vol. 1*, 16–24.

49. Ibid., 25–26.

50. Ibid., 28; emphasis in original.

51. Ibid., 30–38.

52. Ibid., 42–48, 58–64.

53. Ibid., 65; what follows, pp. 67–68.

54. Ibid., 69–77; quotes on pp. 73, 77.

55. Ibid., 79; CCI, United States Public Health Service, and National Health Forum, *Conference on preventive aspects of chronic disease: proceedings March 12–14, 1951, Chicago* (Baltimore, 1951), 10–11.

56. CCI, *Chronic illness in the United States: vol. 1*, 90.

57. Ibid., 90–101.

58. CCI, *Chronic illness in the United States: vol. 2*, 5.

59. Quotes in ibid., 423, 425, 427 respectively.

60. Quotes in ibid., 15, 423.

61. Ibid., 17–21.

62. Ibid., 424.

63. Ibid., 8–20.

64. Ibid., 428.

65. Quotes in paragraph, ibid., 239, 430.

66. Ibid., 431–33.

67. Ibid., 24–25.

68. Ibid., 439.

69. Ibid., 440.

70. Ibid., 441.

71. Ibid., 442.

72. Ibid., 426, 441, 443, 417–18.

73. The discussion below is based on ibid., 395–99, 417–20; "Recommendation of Commission on Chronic Illness re extension of OASI," CMS 2404, 1964, AMA archives, and "Minority report on cash disability benefits under OASI," CMS 2288, 1964, AMA archives. I am grateful to Dr. Carla Keirns, who found and provided me with these documents.

74. Quotes in this paragraph, *Chronic illness in the United States: vol. 2*, 395, 397, 399, respectively.

75. Ibid., 417–19; "The development of the disability program old-age survivors insurance, 1935–74," in House Ways and Means Committee, *Committee staff report on the disability insurance program*, July 1974, 113, www.ssa.gov/history/pdf/dibreport.pdf.

76. CCI, *Chronic illness in the United States: vol. 4, chronic care in a large city* (Cambridge, MA, 1959); CCI, *Chronic illness in the United States: vol. 3, chronic disease in a rural area, reported by Ray E. Trussell and Jack Elinson* (Cambridge, MA, 1957).

77. CCI, "Minutes Meeting of Board of Directors, November 20, 1953," Ernst Boas Papers, American Philosophical Society.

78. CCI, *Chronic illness in the United States: vol. 4*, xv.

79. CCI, *Chronic illness in the United States: vol. 3*, 12.

80. CCI, *Chronic illness in the United States: vol. 4*, xi; *Chronic illness in the United States: vol. 3*, 16

81. CCI, *Chronic illness in the United States: vol. 4*, xvi–xvii.

82. Ibid., 362–64.

83. E.g., D. E. Krueger, "Measurement of prevalence of chronic disease by household interviews and clinical evaluations," *AJPH* 47 (1957); J. Elinson and R. E. Trussell, "Some factors relating to degree of correspondence for diagnostic information as obtained by household interviews and clinical examinations," *AJPH* 47 (1957).

84. M. Bright, "A follow-up study of the Commission on Chronic Illness Morbidity Survey in Baltimore," *J Chron Dis* 20 (1967): 707–16, 717–29; *J Chron Dis* 21 (1969): 749–59.

85. L. W. Mayo et al., "Chronic illness: National Health Forum," *PHR* 71 (1956): 675–96.

86. "Chronic ills study urged by president," *NYT*, March 21, 1956.

87. "The Joseph Earle Moore Clinic," June 20, 1958, in The Victor A. McKusick Papers, Profiles in Science, NLM, http://profiles.nlm.nih.gov/ps/retrieve/ResourceMetadata/ JQBBBY.

88. A. M. Lilienfeld and A. J. Gifford, *Chronic diseases and public health* (Baltimore, 1966).

89. United States, National Health Survey, *Duration of limitation of activity due to chronic conditions, United States, July 1959–June 1960* (Washington, DC, 1962).

90. A. P. Thomson, "Care of chronics in the USA," *BMJ* 2 (1958): 94.

91. Canada, Department of National Health and Welfare and Dominion Bureau of Statistics, *Illness and health care in Canada: Canadian Sickness Survey, 1950–1951* (Ottawa, 1960); K. C. Charron, "Chronic disease in the Canadian Hospital Program," *Canadian Journal of Public Health* 48 (1957): 405–12; idem, "The magnitude of chronic disease in Canada," *Canadian Journal of Public Health* 52 (1961): 273.

92. CCI, *Chronic illness in the United States: vol. 1*, 311.

93. L. A. Scheele, "Progress in prevention of chronic illness, 1949–1956," *JAMA* 160 (1956): 1114.

CHAPTER SIX: Long-Term Care

1. D. Vinyard, "The Senate special committee on the aging," *The Gerontologist* 12 (1972): 298–306.

2. American Medical Association, Committee on Aging, *Report on Conferences on Aging and Long-Term Care: Oklahoma, October 15–16, 1964 and Chicago, Illinois, February 4–5, 1965* (Chicago, 1965), 1–2.

3. A. E. Benjamin, "An historical perspective on home care policy," *Milbank Q* 71 (1993): 139.

4. "West dedicates hospital: San Francisco center modeled after Montefiore here," *NYT*, April 24, 1950.

5. N. Tandon, "Roosevelt Island historical walk: Bird S. Coler Hospital," www.correction history.org/rooseveltisland/html/rooseveltislandtour_coler.html. Accessed Dec. 18, 2012.

6. "The State Chronic Disease Hospital in Boston," *AJPH* 42 (1952): 1469–70; H. T. Phillips, "The intermediate hospital," *N Eng J Med* 276 (1967): 1352–54.

7. In 1961 the hospital's director announced the construction of a new building for "longer-than-average" periods of care in order to free hospital beds in the main building for acutely ill patients. "Montefiore to get convalescent unit," *NYT*, Jan. 16, 1961.

8. R. Stevens, *In sickness and in wealth: American hospitals in the twentieth century (updated edition)* (Baltimore, 1999), 218–24.

9. US Department of Health, Education, and Welfare and Public Health Service, *Hill-Burton program: progress report (summary tables only) July 1, 1947–June 30 1974* (Rockville, MD, 1974), table 9; US Department of Health, Education, and Welfare and Public Health Service, *Hill-Burton program: progress report 1947–1970* (Rockville, MD, 1970), "Introduction"; Stevens, *In sickness and in wealth,* 219.

10. "Nation has 450,000 nursing home beds, PHS survey shows," *Modern Hospital* 84 (1955): 194.

11. US Department of Health, Education, and Welfare and Public Health Service, Division of Hospital and Medical Facilities, and Program Evaluation and Reports Branch, *Hill-Burton program: progress report July 1, 1947–June 30, 1963* (Washington, DC, 1963), 51–55.

12. US Department of Health, Education, and Welfare and Public Health Service, *Hill-Burton program: July 1, 1947–June 30 1974,* table 11.

13. Ibid., tables 11 and 17; "The nation's hospitals: a statistical profile," *Hospitals* 44 (1970): 463.

14. D. N. Finch, "A study of the need for a chronic-disease unit in a general hospital: submitted . . . in partial fulfillment . . . master of hospital administration," University of Michigan, School of Business Administration, 1958; A. Aponte and G. L. Warden, "A design of a method for evaluating a chronic care unit in a general hospital: submitted . . . in partial fulfillment . . . for the degree of master of hospital administration," University of Michigan, School of Business Administration, 1961.

15. D. M. Fox, "Policy and epidemiology: financing health services for the chronically ill and disabled, 1930–1990," *Milbank Q* 67 suppl., 2 pt. 2 (1989): 274.

16. E.g., D. C. Wegmiller, "Extended care," *Hospitals* 41 (1967): 63–66.

17. D. C. Wegmiller, "Long-term care," *Hospitals* 42 (1968): 93–96; H. F. Froeb, S. Dobson, and G. O. Shecter, "An inhalation therapy program for a long-term care institution," *Hospitals* 41 (1967): 76–86.

18. J. R. Norris, "The acute hospital–chronic hospital affiliation: a comparison of length of stay of chronic patients in affiliated and non-affiliated acute hospitals," *Medical Care* 9 (1971): 479–86.

19. W. J. Wentz and G. R. Wonnacott, "Part-time therapist, trained assistant fill occupational therapy gaps," *Hospitals* 41 (1967): 88–95.

20. R. Beckman, "The therapeutic corridor," *Hospitals* 45 (1971): 71–72, 76, 80.

21. W. C. Christenson, "Health planning—continuity of care," *Hospitals* 44 (1970): 21–24.

22. G. E. Ayers and S. P. Mahan, "A sheltered workshop meets authentic needs of the chronically disabled," *Hospitals* 41 (1967): 103–8.

23. H. J. Mast, "ECF's and hospitals," *Hospitals* 44 (1970): 64–67; A. R. Hanna, "A plan for continuing patient care," *Hospitals* 43 (1969): 55–58, 94.

24. "Washington Report—House Committee moves swiftly on Hill-Burton Bill," *Hospitals* 43 (1969): 26.

25. W. Slabodnick, "Profile of a hospital-operated nursing home," *Hospitals* 41 (1967): 70–82; E. L. Forbes, "Hospital-operated retirement residence offers unique advantages," *Hospitals* 41 (1967): 61–64.

26. M. Kaplan, "Experience of a chronic unit affiliated with a general hospital," *J Chron Dis* 19 (1966): 154. Also www.mossrehab.com/About-MossRehab/history-of-moss-rehab .html. Accessed Dec. 18, 2012.

27. M. W. Elliott, "A satellite unit," *Hospitals* 44 (1970): 65–68.

28. Wegmiller, "Long-term care," 94.

29. "The nation's hospitals: a statistical profile," *Hospitals* 44, pt. 2 (1970): 463–68, 493–95.

30. Wegmiller "Long-term care," 95.

31. American Hospital Association, *Hospital statistics: data from the AHA 1979 Annual Survey* (Chicago, 1980), table 12B, p. 204; American Hospital Association, *American Hospital Association hospital statistics, 1990–91 edition; data compiled from the AHA 1989 Annual Survey of Hospitals* (Chicago, 1990), table 12B, p. 215.

32. H. J. Russell and M. V. Wells, "Care beyond acute general hospital," *Hospitals* 44 (1970): 56–59; "The nation's hospitals: a statistical profile," 463–68.

33. D. H. Rhinelander, "Medicare challenges benefits at chronic disease hospitals," *Hartford Courant*, Feb. 5, 1977; A. P. Ruskin, "Whither the chronic disease hospital?" *JAMA* 234 (1975): 709–10.

34. M. Holstein and T. R. Cole, "The evolution of long-term care in America," in *The future of long-term care: social and policy issues*, ed. R. H. Binstock, L. E. Cluff, and O. Von Mering (Baltimore, 1996), 29–34, 37. See also E. Schell, "The origins of geriatric nursing: the chronically ill elderly in almshouses and nursing homes, 1900–1950," *Nursing History Revue* 1 (1993): 203–16.

35. J. Solon, D. W. Roberts, D. E. Krueger, and A.M. Baney, *Nursing homes, their patients and their care: a study of nursing homes and similar long-term care facilities in 13 states*, Public Health Monograph no. 46 (Washington, DC, 1957), 13.

36. Ibid., 45; also see 5–6.

37. E. Eagle, "Nursing homes and related facilities: a review of the literature," *PHR* 83 (1968): 673–84.

38. F. M. Foote, "Progress in nursing home care," *JAMA* 202 (1967): 296–98; Eagle, "Nursing homes," 680.

39. Eagle, "Nursing homes," 674–75.

40. "News—number of ECF beds passes million mark," *Hospitals* 43 (1969): 165.

41. Holstein and Cole, "Evolution of long-term care," 34–40.

42. J. A. Solon and L. F. Greenawalt, "Physicians' participation in nursing homes," *Medical Care* 12 (1974): 487, figure 1.

43. "News—ANHA calls Medicare 'hoax' on elderly, withdraws support of Extended Care Program," *Hospitals* 45 (1971): 130–31.

44. J. Feder and W. Scanlon, "The underused benefit: Medicare's coverage of nursing home care," *Milbank Q* 60 (1982): 604–32, especially 604–7.

45. Eagle, "Nursing homes," 679–81.

46. B. C. Vladeck, *Unloving care: the nursing home tragedy* (New York, 1980); J. Takeuchi,

R. Burke, and M. McGeary, *Improving the quality of care in nursing homes—Institute of Medicine Committee on Nursing Home Regulation* (Washington, DC, 1986).

47. B. C. Vladeck, "Unloving care revisited: the persistence of culture," *Journal of Social Work in Long-Term Care* 2 (2003): 1–9; see also A. L. Jones, L. L. Dwyer, A. R. Bercovitz, and G. W. Strahan, "The National Nursing Home Survey: 2004 overview," in *Vital and health statistics, Series 13, data from the National Health Survey* 167 (2009): 1–7.

48. R. N. Knollmueller, "Funding home care in a climate of cost containment," *Public Health Nursing* 1 (1984): 16.

49. E.g., P. Sloane, D. R. Redding, and L. Wittlin, "Longest-term placement problems in an acute care hospital," *J Chron Dis* 34 (1981): 285–90; W. G. Weissert and C. M. Cready, "Determinants of hospital-to-nursing home placement delays: a pilot study," *Health Services Research* 23 (1988): 619.

50. Holstein and Cole, "Evolution of long-term care," 4.

51. K. Buhler-Wilkerson, "Care of the chronically ill at home: an unresolved dilemma in health policy for the United States," *Milbank Q* 85 (2007): 611–39.

52. E. Boris and J. Klein, *Caring for America: home health workers in the shadow of the welfare state* (New York, 2012).

53. Commonwealth Fund, Frode Jensen, H. G. Weiskotten, and M. A. Thomas, *Medical care of the discharged hospital patient* (New York, 1944); CCI, *The Commission on Chronic Illness; origin, structure, objectives, staffing and financing* (Chicago, 1949), 7.

54. E.g., E. M. Bluestone, *Home care: origin, organization and present status of the extramural program of Montefiore Hospital* (New York, 1949); idem, "The principles and practice of home care," *JAMA* 155 (1954): 1379–82.

55. CCI, *Chronic illness in the United States: vol. 2, care of the long-term patient* (Cambridge, MA, 1957), 66.

56. Ibid., 67–69, 81–83.

57. US Public Health Service and CCI, *A study of selected home care programs; a joint project of the Public Health Service and the Commission on Chronic Illness* (Washington, DC, 1955); CCI, *Chronic illness in the United States: vol. 2*, 67–78.

58. CCI, *Chronic illness in the United States: vol. 2*, 78.

59. Buhler-Wilkerson, "Care of the chronically ill," 620.

60. Benjamin, "Historical perspective on home care," 129–66.

61. H. J. Russell and M. V. Wells, "Care beyond acute general hospitals," *Hospitals* 44 (1970): 56–59; E. L. Harmon, "Third-party payment increases utilization of home care services," *Hospitals* 42 (1968): 68–72.

62. V. K. Volk and R.R. Manty, "Day care program aids patients in transition to home care," *Hospitals* 41 (1967): 76–82.

63. Russell and Wells, "Care beyond acute general hospitals," 56–59; Harmon, "Third-party payment," 68–72.

64. M. F. Rappaport, "Community care homes," *Hospitals* 44 (1970): 57–59.

65. R. Morris and E. Harris, "Home health services in Massachusetts, 1971: their role in care of the long-term sick," *AJPH* 62 (1972): 1088–93; K. Ricker-Smith and B. Trager, "In-home health services in California: some lessons for national health insurance," *Medical Care* 16 (1978): 173–90.

66. Benjamin, "Historical perspective on home care," 139–51, 159–60; Buhler-Wilkerson, "Care of the chronically ill," 624–25.

67. Buhler-Wilkerson, "Care of the chronically ill," 625; Benjamin, "Historical perspective on home care," 147.

68. Buhler-Wilkerson, "Care of the chronically ill," 626.

69. American Hospital Association, *Hospital statistics 1979 Annual Survey*, table 12B, p. 204; idem, *Hospital statistics 1989 Annual Survey*, table 12B, p. 215.

70. Buhler-Wilkerson, "Care of the chronically ill," 612–13.

71. "Construction notes," *Hospitals* 43 (1969): 38.

72. B. C. Horstman, "Hospitals and nursing homes cooperate in rehabilitation workshops," *Hospitals* 42 (1968): 68–72, 76.

73. B. M. Dornblaser and E. J. Rising, "Hospital-based extended care part 1: conducting an ADL rehabilitation program," *Hospitals* 42 (1968): 68–70, 79–82.

74. C. M. Wylie, "Early rehabilitation promises greater improvement to stroke patients," *Hospitals* 42 (1968): 100–104.

75. E. D. Berkowitz, "The federal government and the emergence of rehabilitation medicine," *Historian* 43 (1981): 530–45; R. Verville, *War, politics, and philanthropy: the history of rehabilitation medicine* (Lanham, MD, 2009); H. A. Rusk, "Rehabilitation," *JAMA* 140 (1949): 286–92.

76. C. E. Caniff, "Rehabilitation centers—characteristics and trends," *Conference on rehabilitation concepts held at University of Pennsylvania, Philadelphia, October 17–18, 1962* (Chicago, 1963), 100.

77. Verville, *War, politics, and philanthropy*, 171; S. S. Lee and M. I. May, "Three year appraisal of a hospital-home affiliation," *Hospitals* 42 (1968): 103–6.

78. I. J. Brightman, "The proposed federal legislation for independent living rehabilitation," *AJPH* 51 (1961): 753–59; Verville, *War, politics, and philanthropy*, 206.

79. M. Berkowitz, *An evaluation of policy-related rehabilitation research* (New York, 1975).

80. H. R. Kelman, "An experiment in the rehabilitation of nursing home patients," *PHR* 77 (1962): 356–66; W. E. Park and M. I. Moe, "Rehabilitation care in nursing homes," *PHR* 75 (1960): 605–13.

81. L. Z. Rubenstein et al., "Effectiveness of a geriatric evaluation unit: a randomized clinical trial," *N Eng J Med* 311 (1984): 1664–70.

82. G. Becker, "Age bias in stroke rehabilitation: effects on adult status," *Journal of Aging Studies* 8 (1994): 271–90; M. G. Ory and T. F. Williams, "Rehabilitation: small goals, sustained interventions," *Ann AAPSS* 503 (1989): 60–71.

83. American Hospital Association, Hospital Statistics, 1975, 1979, 1980, 1990 editions, Chicago Illinois. For the latter two editions the data is in table 12A.

84. R. Mullner, F. Nuzum, and D. Matthews, "Inpatient medical rehabilitation: results of the 1981 Survey of Hospitals and Units," *Archives of Physical and Medical Rehabilitation* 64 (1983): 354–57.

85. For an expression of this ideal by a pioneer of the field, see H. A. Rusk, "Rehabilitation belongs in the general hospital: otherwise, a major attack on the problems of chronic disability cannot succeed," *AJN* 62 (1962): 62–63.

86. A. Walker, "Rehabilitation: micro-market or major partner in healthcare's future?" *Journal of Rehabilitation* 54 (1988): 19–22.

87. Verville, *War, politics, and philanthropy*, 220–41.

88. S. Intagliata and R. Hollander, "The 3-hour therapy criterion: a challenge for rehabilitation facilities," *American Journal of Occupational Therapy* 41(1987): 297–304.

89. L Chan et al., "The effect of Medicare's payment system for rehabilitation hospitals on length of stay, charges, and total payments," *N Eng J Med* 337 (1997): 978–85.

90. A. W. Heinemann, J. Billeter, and H. B. Betts, "Prospective payment for acute care: impact on rehabilitation hospitals," *Archives of Physical Medicine & Rehabilitation* 69 (1988): 614–18; J. L. Melvin, "The 20th Walter J. Zeiter lecture: trends in delivery and funding of postacute care," *Archives of Physical Medicine & Rehabilitation* 69 (1988): 163–66.

91. G. Gritzer and A. Arluke, *The making of rehabilitation: a political economy of medical specialization, 1890–1980* (Berkeley, CA, 1989), 148.

92. E.g., W. M. Fowler, "Viability of physical medicine and rehabilitation in the 1980s," *Archives of Physical Medicine & Rehabilitation* 63 (1982): 1–5; A. L Caplan, D. Callahan, and J. Haas, "Ethical and policy issues in rehabilitation medicine," *Hastings Centre Report* 17, suppl. 1 (1987): 20.

93. American Board of Medical Specialties 2012 Certification (Chicago, 2013); J. D. Leigh, D. Tancredi, A. Jerant, and R. L. Kravitz, "Physician wages across specialties: informing the physician reimbursement debate," *Archives of Internal Medicine* 170 (2010): 1728–34.

CHAPTER SEVEN: Public Health and Prevention

1. L. Breslow, *A life in public health: an insider's retrospective* (New York, 2004), 64–65; W. L. Halverson, L. Breslow, and M. H. Merrill, "Chronic disease—the Chronic Disease Study of the California Department of Public Health," *AJPH* 39 (1949): 594.

2. E.g., M. Terris, "The changing face of public health," *AJPH* 49 (1959): 1113–19; H. E. Hilleboe, "Public health and medicine at the crossroads," *PHR* 67 (1952): 767–71.

3. American Public Health Association, Committee on Chronic Disease and Rehabilitation, *Chronic disease and rehabilitation: a program guide for state and local health agencies* (New York, 1960), 10.

4. W. David, "Chronic respiratory diseases—the new look in the public health service," *AJPH* 57 (1967): 1357–62.

5. J. B. Graber, "Public health service programs on aging," *PHR* 79 (1964): 577–81; idem, "Findings and implications of a nationwide program review of resources available to meet the health needs of the aging and aged," *Gerontologist* 6 (1966): 191–200.

6. A. J. McDowell, "U.S. National Health Examination Survey," *PHR* 80 (1965): 941–48.

7. T. D. Woolsey, P. S. Lawrence, and E. Balamuth, "An evaluation of chronic disease prevalence data from the health interview survey," *AJPH* 52 (1962): 1631–37; J. V. Neel et al., *Genetics and the epidemiology of chronic diseases; a symposium, June 17–19, 1963, Ann Arbor, Michigan*, Public Health Service publication no. 1163 (Washington, DC, 1965), 196; David, "Chronic respiratory diseases," 1361; Chronic Respiratory Diseases Control Pro-

gram and National Center for Chronic Disease, *Current research in chronic airways obstruction: Ninth Aspen Emphysema Conference, Aspen, Colorado, June 9–12, 1966*, Public Health Service publication no. 1717 (Washington, DC, 1968).

8. H. W. Miller, *Plan and operation of the National Health and Nutrition Examination Survey, 1971–1973*, DHEW Publication no. (PHS) 79-1310 (Hyattsville, MD, 1973).

9. David, "Chronic respiratory diseases," 1360.

10. B. Kenadjian, "Appropriate types of federal grants for state and community health services," *PHR* 81 (1966): 815–20.

11. L. Robins, "The impact of decategorizing federal programs: before and after 314(d)," *AJPH* 62 (1972): 24–29.

12. S. P. Strickland, *Politics, science, and dread disease: a short history of United States medical research policy* (Cambridge, MA, 1972), ch. 3.

13. A short history of the National Institutes of Health, http://history.nih.gov/exhibits/history/docs/page_06.html, accessed Aug. 11, 2012; Strickland, *Politics, science, and dread disease*, 75, 182.

14. Strickland, *Politics, science, and dread disease*, 75, 182; B. S. Park, "Disease categories and scientific disciplines: reorganizing the NIH Intramural Program, 1945–1960," in *Biomedicine in the twentieth century: practices, policies, and politics*, ed. C. Hannaway (Amsterdam, 2008), 34; A.N.H. Creager, "Mobilizing biomedicine: virus research between lay health organizations and the U.S. Federal Government," in ibid., 182.

15. W. H. Sebrell and C. V. Kidd, "Administration of research in the National Institutes of Health," *Scientific Monthly* 74 (1952): 154.

16. W. H. Sebrell, "Nutrition research; potentialities in chronic disease," *PHR* 68 (1953): 737–41.

17. *Chemical Engineering News* 31 (1953): 2990–92.

18. Transcript Longines Chronoscope, interview with C. J. Van Slyke, 10/30/53, http://ahiv.alexanderstreet.com/View/Transcript/528181. Accessed Oct. 11, 2010.

19. E.g., V. Quirke, "From evidence to market: Alfred Spinks' 1953 survey of fields for pharmacological research, and the origins of ICI's Cardiovascular Programme," in *Medicine, the market and the mass media: producing health in the twentieth century*, ed. V. Berridge and K. Loughlin (London, 2005), 146–71; J. A. Greene, *Prescribing by numbers: drugs and the definition of disease* (Baltimore, 2007).

20. "Institute of Chronic Disease set at UCLA," *Los Angeles Times*, Aug. 9, 1967.

21. H. M. Thomas, "Dr. Joseph Earl Moore," *Transactions of the American Clinical Climatological Association* 71 (1960): xlix–l.

22. D. E. Chubin and K. E. Studer, "The politics of cancer," *Theory and Society* 6 (1978): 64; R. A. Rettig, "Reflections on 'the Cancer Crusade,'" Rand Corporation Paper, P-6257 (Santa Monica, CA, 1978), 19.

23. A. N. Barker and H. Jordan, "Legislative history of the National Cancer Program," in NCBI Resources, www.ncbi.nlm.nih.gov/books/NBK13873/. Accessed Oct. 30, 2012.

24. B. C. Schmidt, "Five years into the National Cancer Program: retrospective perspectives—the National Cancer Act of 1971," *Yale J Biol Med* 50 (1977): 240.

25. R. C. Brownson and F. S. Bright, "Chronic disease control in public health practice: looking back and moving forward," *PHR* 119 (2004): 233.

26. CCI, *Chronic illness in the United States: vol. 1, prevention of chronic illness* (Cambridge, MA, 1957), 100–101.

27. "The issues in chronic disease control: a conference report," *PHR* 75 (1960): 827–34; L. Alvin and P. Nicholas, *Partnership in learning, an historical report, 1960–1966* (Washington, DC, 1967).

28. G. D. Carlyle Thompson, "Health department services in chronic disease," *AJPH* 46 (1956): 1543–46; J. N. Muller and E. B. Kovar, "Chronic disease services in local health departments—report of a survey," *AJPH* 47 (1957): 352–62; A. L. Chapman and D. Bergsma, "State grants for local projects in chronic illness control," *PHR* 71 (1956): 337–39.

29. C. A. Miller et al., "Statutory authorizations for the work of local health departments," *AJPH* 67 (1977): 940–45.

30. Ibid.; American Public Health Association, Program Area Committee on Chronic Disease and Rehabilitation, *Chronic disease and rehabilitation: a program guide for state and local health agencies* (New York, 1960), 20.

31. American Public Health Association, Program Area Committee on Chronic Disease and Rehabilitation, *Chronic disease and rehabilitation*, 20; C. A. Miller et al., "A survey of local public health departments and their directors," *AJPH* 67 (1977): 931–39.

32. Breslow, *A life in public health*, 105–6; Halverson, Breslow, and Merrill, "Chronic disease, California"; L. Breslow and H. W. Mooney, "The California Morbidity Survey—a progress report," *California Medicine* 84 (1956): 95–97; L. Breslow, "Uses and limitations of the California Health Survey for studying the epidemiology of chronic disease," *AJPH* 47 (1957): 168–72; L. Breslow, N. Ott, and V. Chin, "California's chronic disease activities," *PHR* 71 (1956): 453–58.

33. L. Breslow, "Periodic health examinations and multiple screening," *AJPH* 49 (1959): 1148–56; "'Multiphasic testing, 1971,' a socio-economic report of the Bureau of Research and Planning," *California Medicine* 114 (1971): 71–81.

34. L. Breslow, "From disease prevention to health promotion," *JAMA* 281 (1999): 1030–33; Breslow, *A life in public health*, 84–86.

35. New York Health Preparedness Commission, *Planning for the care of the chronically ill in New York State; some medical-social and institutional aspects* (Albany, 1947).

36. W. L. Oliver, "Multiple screening tests for chronic diseases," *New York State Journal of Medicine* 67 (1967): 302–8; I. J. Brightman, "Chronic disease programs—more sanity and more service," *AJPH* 54 (1964): 845–47.

37. P. M. Densen, G. James, and E. Cohart, "Research, program planning, and evaluation," *PHR* 81 (1966): 49–56; E. R. Schlesinger, A. M. Bahlke, and A. R. Cohen, "Development of a state-wide program for the care of children with long-term illness," *AJPH* 55 (1965): 973–77.

38. David, "Chronic respiratory diseases"; J. H. Ludwig and B. J. Steigerwald, "Research in air pollution: current trends," *AJPH* 55 (1965): 1082–92.

39. O. H. Wood, D. H. Hollander, and J. A. McCallum, "The Maryland state chronic disease hospitals," *Maryland Medical Journal* 7 (1958): 672–78; P. F. Prather, "Maryland's approach to integrated medical and hospital services," *AJPH* 51 (1961): 1137–43.

40. P. S. Lawrence and C. Tibbitts, "Recent long-term morbidity studies in Hagers-

town, Md.," *AJPH* 41 (1951): 101–7; J. Downes, "Findings of the Study of Chronic Disease in the Eastern Health District of Baltimore," *Milbank Q* 22 (1944): 337–51; M. Bright, "A follow-up study of the Commission on Chronic Illness morbidity survey in Baltimore," *J Chronic Dis* 20 (1967): 707–16, 717–29; L. Kuller and S. Tonascia, "A follow-up study of the Commission on Chronic Illness morbidity survey in Baltimore," *J Chronic Dis* 24 (1971): 111–24.

41. M. Grant and W. E. Paupe, "Countywide screening programs for chronic disease," *PHR* 78 (1963): 767–71; M. Grant, "Screening for chronic disease with a mobile health unit," *PHR* 80 (1965): 633–36.

42. A. L. Chapman and D. Bergsma, "State grants for local projects in chronic illness control," *PHR* 71 (1956): 338; R. P. Kandle, "Experience with a state chronic illness law," *AJPH* 54 (1964): 103–7.

43. Kandle, "Experience with a state chronic illness law," 104.

44. E. K. Caso and H. T. Phillips, "Small-grants projects in Massachusetts for the chronically ill and aged," *PHR* 81 (1966): 471–77.

45. J. W. Runyan Jr., W. E. Phillips, O. Herring, and L. Campbell, "A program for the care of patients with chronic diseases," *JAMA* 211 (1970): 476–79.

46. E. M. Holmes Jr. and P. W. Bowden, "Current experience in multiphasic health examinations; experience at Richmond," *AJPH* 41 (1951): 640–41; L. Blumenkranz and F. J. Spencer, "Patients with chronic disease in Richmond's home care program: selected data, December 1965," *PHR* 83 (1968): 75–80.

47. Muller and Kovar, "Chronic disease services in local health departments," 361.

48. Centers for Disease Control and Association of State Territorial Health Officials, *First National Conference on Chronic Disease Prevention and Control: Identifying Effective Strategies: September 9–11, 1986, Atlanta: conference summary* (Atlanta, GA, 1987), 5–6.

49. P. Q. Peterson, "The health department's responsibility in chronic disease programs," *AJPH* 50 (1960): 134–39; "The issues in chronic disease control: a conference report," *PHR* 75 (1960): 827–34.

50. P.K.J. Han, "Historical changes in the objectives of the periodic health examination," *Annals of Internal Medicine* 127 (1997): 913.

51. A. Morabia and F. F. Zhang, "History of medical screening: from concepts to action," *Postgraduate Medical Journal* 80 (2004): 467.

52. Ibid., 464, 465–66: A. B. Kurlander and B. E. Carroll, "Case finding through multiple screening," *PHR* 68 (1953): 1035; R. H. Fetz and L. M. Petrie, "The Anthrone blood sugar method adapted to diabetes case finding in a multiple screening program," *PHR* 65 (1950): 1709–18.

53. L. Breslow, "A historical review of multiphasic screening," *Preventive Medicine* 2 (1973): 177.

54. C. K. Canelo, D. M. Bissell, H. Abrams, and L. Breslow, "A multiphasic screening survey in San Jose," *California Medicine* 71 (1949): 409–11; V. A. Getting and H. L. Lombard, "The evaluation of pilot clinics: the mass screening or health-protection plan," *N Eng J Med* 247 (1952): 460–65.

55. W. L. Laurence, "Multiphase tests show hidden ills," *NYT*, Nov. 4, 1950.

56. Quoted in L. Breslow, "Periodic health examinations and multiple screening," *AJPH* 49 (1959): 1152.

57. Ibid.

58. Ibid., 1148.

59. Ibid., 1153–55. For specific examples, see M. Grant, "Screening for chronic disease with a mobile health unit," *PHR* 80 (1965): 633–36; J. J. Adler, C. M. Bloss Jr., and K. T. Mosley, "The Oklahoma State Health Department Mobile Multiphasic Screening Program for Chronic Disease," *AJPH* 56 (1966): 918–25.

60. M. F. Collen, "Introduction to Multiphasic Health Testing Forum," *Preventive Medicine* 2 (1973): 175–76; Han, "Historical changes," 914.

61. J. Houston, "Computer medical exam: impersonal but practical," *Chicago Tribune*, Sept. 2, 1973.

62. J. Cook, "Thousands take heart tests," *NYT*, Nov. 10, 1974. On the relationship between screening and risk factors, see David Armstrong, "Screening: mapping medicine's temporal spaces," *Sociology of Health & Illness* 34 (2012): 177–93.

63. Han, "Historical changes," 914; G. F. Friedman, M. F. Collen, and B. H. Fireman, "Multiphasic health checkup evaluation: a 16-year follow-up," *J Chron Dis* 39 (1986): 1434–36. On false positives and negatives, see S. J. Reiser, "The emergence of the concept of screening for disease," *Milbank Q* 56 (1978): 416–18.

64. M. J. Casper and A. E. Clarke, "Making the pap smear into the right tool for the job: cervical cancer screening in the USA, circa 1940–95," *Social Studies of Science* 28 (1998): 255–90; Ilana Löwy, *Preventive strikes: women, precancer, and prophylactic surgery* (Baltimore, 2009), 118–65; R. C. Brownson and F. S. Bright, "Chronic disease control in public health practice: looking back and moving forward," *PHR* 119 (2004): 230–38.

65. K. M. Chacko and R. J. Anderson, "The annual physical examination: important or time to abandon," *American Journal of Medicine* 120 (2007): 581–83. See also Canadian Task Force on the Periodic Health Exam, "The periodic health exam," *Canadian Medical Association Journal* 121 (1979): 1193–1254.

66. Han, "Historical changes."

67. L. Breslow and J. R. Hochstim, "Sociocultural aspects of cervical cytology in Alameda County, Calif.," *PHR* 79 (1964): 107–12.

68. E.g., L. Breslow, "Medical care and health education: some new relationships," *PHR* 83 (1968): 791–95; H. Gales, "The Community Health Education Project: bridging the gap," *AJPH* 60 (1970): 322–27.

69. M. A. Guinta and J. P. Allegrante, "The President's Committee on Health Education: a 20-year retrospective on its politics and policy impact," *AJPH* 82 (1992): 1033–41.

70. American Public Health Association, *A model for planning patient education: an essential component of health care* (Washington, DC, 1975). Quote is from the online abstract in http://catalogue.nla.gov.au/Record/5312458, accessed Dec. 27, 2012.

71. W. G. Rothstein, *Public health and the risk factor: a history of an uneven medical revolution* (Rochester NY, 2003); L. Berlivet, "'Association or causation?' the debate on the scientific status of risk factor epidemiology, 1947–c. 1965," *Clio Medica* 75 (2005): 39–74;

É. Giroux, "The Framingham study and the constitution of a restrictive concept of risk factor," *Soc Hist Med* 26 (2013): 94–112.

72. L. Breslow, "Risk factor intervention for health maintenance," *Science* 200 (1978): 908–12; "Multiple Risk Factor Intervention Trial; risk factor changes and mortality results," *JAMA* 248 (1982): 1465–77.

73. E.g., L. Breslow, "Medical care and health education," *PHR* 83 (1968): 791–95; H. Leventhal, "Changing attitudes and habits to reduce risk factors in chronic disease," *American Journal of Cardiology* 31 (1973): 571–80; C. I. Cohen and E. J. Cohen, "Sounding board: Health education: panacea, pernicious or pointless?" *N Eng J Med* 299 (1978): 718–20.

74. N. K. Janz and M. H. Becker, "The health belief model: a decade later," *Health Education & Behavior* 11(1984): 1–47. According to Google Scholar, this article has been cited more than four thousand times.

75. D. W. Jones and J. E. Hall, "The National High Blood Pressure Education Program: thirty years and counting," *Hypertension* 39 (2002): 941–42.

76. US Department of Health, Education, and Welfare, *Healthy people: the Surgeon-General's Report on Health* (Washington, DC, 1979); R. M. Kaplan, "Behavioral epidemiology, health promotion, and health services," *Medical Care* 23 (1985): 564–83.

77. S. Havas and B. Walker Jr., "Massachusetts' approach to the prevention of heart disease, cancer, and stroke," *PHR* 101 (1986): 29–39.

78. E.g., American Hospital Association—Center for Health Promotion and CDC Center for Health Promotion and Education Community Program Development Division, *Patient education in hospital-sponsored home health care programs* (Atlanta, GA, 1982).

79. D. Neubauer and R. Pratt, "The second public health revolution: a critical appraisal," *Journal of Health Politics, Policy and Law* 6 (1981): 205–28; M. Minkler, "Health education, health promotion and the Open Society: an historical perspective," *Health Education & Behavior* 16 (1989): 17–30; WHO, *Health promotion: a discussion document of the concept and principles* (Copenhagen, 1984).

80. J. T. O'Connor, "Comprehensive health planning: dreams and realities," *Milbank Q* 52 (1974): 391–413.

81. President's Commission on Heart Disease Cancer and Stroke, *A national program to conquer heart disease, cancer and stroke; report to the President* (Washington, DC, 1964).

82. A. L. Komaroff, "Regional medical programs in search of a mission," *N Eng J Med* 284 (1971): 758–64; NLM, *Briefing book on the history of regional medical programs* (Bethesda, MD, 1991).

83. CDC Center for Health Promotion and Education Community Program Development Division and American Hospital Association Center for Health Promotion, *Health education in health departments* (Chicago, 1981), 9–15.

84. "Progress in chronic disease prevention: chronic disease prevention and control activities—United States, 1989," *MMWR—Center for Disease Control* 40 (1991): 697–700.

85. "Resources and priorities for chronic disease prevention and control, 1994," *MMWR—Center for Disease Control* 46(1997): 286–87.

86. H. I. Meissner, L. Bergner, and K. M. Marconi, "Developing cancer control capacity in state and local public health agencies," *PHR* 107 (1992): 15–23.

87. W. H. Foege, "The changing priorities of the Center for Disease Control," *PHR* 93 (1978): 616–21; R. C. Brownson and F. S. Bright, "Chronic disease control in public health practice: looking back and moving forward," *PHR* 119 (2004): 233.

88. D. E. Nelson, E. Powell-Griner, M. Town, and M. G Kovar, "A comparison of national estimates from the National Health Interview Survey and the Behavioral Risk Factor Surveillance System," *AJPH* 93 (2003): 1335–41; "BRFSS History," www.cdc.gov/brfss/history.htm, accessed Sept. 11, 2012.

89. "Chronic disease home," www.cdc.gov/chronicdisease/about/index.htm#mvp. Accessed Nov. 12, 2012.

90. R. S. Schweiker, "Strategies for disease prevention and health promotion in the Department of Health and Human Services," *PHR* 97 (1982): 197.

91. J. O. Mason, J. P. Koplan, and P. M. Layde, "The prevention and control of chronic diseases: reducing unnecessary deaths and disability—a conference report," *PHR* 102 (1987): 20; emphasis mine.

92. D. M. Fox, *Power and illness: the failure and future of American health policy* (Berkeley, 1993).

CHAPTER EIGHT: Health, Wealth, and the State

1. *Statistical yearbook League of Nations 1930–31*, http://digital.library.northwestern.edu/league/le0267ae.pdf; Population Division, DESA, United Nations: *World Population Ageing 1950–2050*, www.un.org/esa/population/publications/worldageing19502050/countriesorareas.htm.

2. *JAMA* 146 (1951): 1060.

3. US Senate Special Committee on Aging, *Economics of aging: toward a full share in abundance: hearings before the Special Committee on Aging, United Stated Senate—Part—International perspectives, August 25, 1969* (Washington, DC, 1970).

4. M. Lalonde, *A new perspective on the health of Canadians: a working document* (Ottawa, 1974).

5. WHO Regional Office for Europe, *The public health aspects of chronic disease: report of a symposium sponsored by the Regional Office for Europe of the World Health Organization in collaboration with the Government of the Netherlands, Amsterdam, 30 September–5 October* (Copenhagen, 1958).

6. WHO, *Role of hospitals in programmes of community health protection: first report of the Expert Committee on Organization of Medical Care*, Technical Report Series no. 122 (Geneva, 1957); WHO, *Role of hospitals in ambulatory and domiciliary medical care: second report of the Expert Committee on Organization of Medical Care*, Technical Report Series no. 176 (Geneva, 1959).

7. WHO, *Expert Committee on Medical Rehabilitation: first report*, Technical Report Series no. 158 (Geneva, 1969); WHO, *Expert Committee on Medical Rehabilitation: second report*, Technical Report Series no. 419 (Geneva, 1958).

8. See L. Breslow, *A life in public health: an insider's retrospective* (New York, 2004), 104–6; B. K. Tones, "Health education and the ideology of health promotion: a review of alternative approaches," *Health Education Research* 1 (1986): 3–12.

9. For an exemplary case, see N. Henckes, "Narratives of change and reform processes:

global and local transactions in French psychiatric hospital reform after the Second World War," *Soc Sci Med* 68 (2009): 511–18.

10. Z. Pisa, "The World Health Organization's heart control programme," in *Chronic diseases, WHO Regional Office for Europe* (Copenhagen, 1973), 2.

CHAPTER NINE: Alternative Paths in the United Kingdom

1. G. Newman, *An outline of the practice of preventive medicine: a memorandum addressed to the Minister of Health* (London, 1919); Lord Dawson of Penn, *Interim report on the future provisions of medical and allied services, United Kingdom Ministry of Health Consultative Council on Medical Allied Services* (London, 1920).

2. Great Britain, *On the state of the public health—annual report of the Chief Medical Officer of the Ministry of Health for 1927* (London, 1928), 104.

3. W. Willcox, "Foreword," *Annals of Rheumatic Diseases* 1 (1939): 1–4.

4. M. Denham, "The surveys of the Birmingham sick hospitals, 1948–1960s," *Soc Hist Med* 19 (2006): 279–80.

5. On the medical functions of the Poor Law, see M. A. Crowther, "Paupers or patients? Obstacles to professionalization in the Poor Law Medical Service Before 1914," *J Hist Med Allied Sci* 39 (1984): 33–54; "The medical recommendations of the Poor Law commission," *Public Health* 22 (1908): 269–71.

6. M. Gorsky, J. Mohan, and T. Willis, *Mutualism and health care: British hospital contributory schemes in the twentieth century* (Manchester, 2006).

7. A. Levene, "Between less eligibility and the NHS: the changing place of Poor Law hospitals in England and Wales, 1929–39," *Twentieth Century British History* 20 (2009): 322–45.

8. A. Levene, M. Powell, and J. Stewart, "The development of municipal general hospitals in English county boroughs in the 1930s," *Medical History* 50 (2006): 3–28, especially 8–12; M. A. Crowther, "From workhouse to NHS hospital in Britain, 1929–1948," in *The Poor Law and after: workhouse hospitals and public welfare*, ed. C. Hillam and J. M. Bone (Liverpool, 1999), 38–48.

9. Levene, "Between less eligibility and the NHS," 323; M. Gorsky, "Creating the Poor Law legacy: institutional care for older people before the welfare state," *Contemporary British History* 26 (2012): 441–65.

10. Levene, "Between less eligibility and the NHS," 332.

11. M. Gorsky. "Local government health services in interwar England: problems of quantification and interpretation," *Bull Hist Med* 85 (2011): 384–412.

12. E.g., *On the state of the public health—annual report 1927*, 232–33, 98–100; Great Britain, *On the state of the public health—annual report of the Chief Medical Officer of the Ministry of Health for 1931* (London, 1932), 89–94.

13. D. Cantor, "Introduction: cancer control and prevention in the twentieth century," *Bull Hist Med* 81 (2007): 1–38; "National radium centres," *BMJ* 1931, 2: 713–17. This publication lacked annual volume numbers until the 1980s and published several volumes annually, numbered 1, 2, or 3. I cite sources from the 1950s and 1960s by date and volume number.

14. Great Britain, *On the state of the public health—annual report of the Chief Medical*

Officer of the Ministry of Health for 1928 (London, 1929), 265; Great Britain, *On the state of the public health—annual report of the Chief Medical Officer of the Ministry of Health for 1929* (London, 1930), 64.

15. J. Welshman, "The medical officer of health in England and Wales, 1900–1974: watchdog or lapdog?" *Journal of Public Health* 19 (1997): 443–50; M. Gorsky, "Public health in interwar England and Wales: did it fail?" *Dynamis* 28 (2008): 175–98.

16. C. Webster, "The elderly and the early National Health Service," in *Life, death, and the elderly: historical perspectives*, ed. M. Pelling and R. M. Smith (London, 1991), 167–93; Gorsky, "Creating the Poor Law legacy"; S. Pickard, "The role of governmentality in the establishment, maintenance and demise of professional jurisdictions: the case of geriatric medicine," *Sociol Health Illn* 32 (2010): 1075.

17. P. Townsend, "Social surveys of old age in Great Britain, 1945–58," *Bull WHO* 21 (1959): 583–91.

18. M. Martin, "Medical knowledge and medical practice: geriatric medicine in the 1950s," *Soc Hist Med* 8 (1995): 443–61.

19. E.g., A. R. Culley, "The care of the aged and infirm and the chronic sick," *Public Health* 60 (1946–47): 102–4.

20. Webster, "The elderly"; Martin, "Medical knowledge"; P. Bridgen, "Hospitals, geriatric medicine, and the long-term care of elderly people 1946–1976," *Soc Hist Med* 14 (2001): 507–23; M. Gorsky, "'To regulate and confirm inequality'? A regional history of geriatric hospitals under the English National Health Service, c. 1948–c. 1975," *Ageing and Society* 33 (2013): 598–625.

21. British Medical Association, *The care and treatment of the elderly and infirm* (London, 1947).

22. In addition to the works cited in note 20, see Pickard, "The role of governmentality"; Gorsky, "Creating the Poor Law legacy"; Denham, "Surveys of the Birmingham chronic sick hospitals."

23. Bridgen, "Hospitals, geriatric medicine," 514. See also C. A. Boucher, *Survey of services available to the chronic sick and elderly, 1954–1955*, Reports on Public Health and Medical Subjects no. 98 (London, 1957); Denham, "Surveys of the Birmingham chronic sick hospitals," 287–88.

24. Bridgen, "Hospitals, geriatric medicine," 516–21.

25. J. Brooks, "Managing the burden: nursing older people in England, 1955–1980," *Nursing Inquiry* 18 (2011): 226.

26. Pickard, "The role of governmentality," 1080; Denham, "Surveys of the Birmingham chronic sick hospitals," 289–91.

27. P. Townsend, *The last refuge: a survey of residential institutions and homes for the aged in England and Wales* (London, 1962); B. Robb, *Sans everything: a case to answer* (London, 1967).

28. R. Klein, "Policy making in the National Health Service," *Political Studies* 22 (1974): 7; Bridgen, "Hospitals, geriatric medicine," 522; J. Lewis, "Older people and the health-social care boundary in the UK: half a century of hidden policy conflict," *Social Policy and Administration* 35 (2001): 343–59.

29. L. J. Donaldson, "1969 revisited: reflections on tomorrow's community physician," *Int J Epid* 30 (2001): 1174.

30. Lewis, "Older people and the health-social care boundary"; for a more detailed version of this argument, see P. Bridgen and J. Lewis, *Elderly people and the boundary between health and social care, 1946–91: whose responsibility? Nuffield Trust grant report* (London, 1999).

31. On Bristol, see Gorsky, "'To regulate and confirm inequality'"; J. Welshman, "Growing old in the city: public health and the elderly in Leicester, 1948–74," *Medical History* 40 (1996): 74–89.

32. A. P. Thomson, "Problems of ageing and chronic sickness," *BMJ* 1949, 2: 243–50, 304–5.

33. For an overview, see Denham, "Surveys of the Birmingham chronic sick hospitals," 284.

34. M. Warren, "Care of the chronic sick," *BMJ* 1943, 2: 822–23.

35. W. Hughes and S. L. Pugmire, "A geriatric hospital service," *Lancet* 259 (1952): 1249–54. I am grateful to Martin Gorsky for calling this tradition to my attention.

36. T. McKeown and C. R. Lowe, "A scheme for the care of the aged and chronic sick," *BMJ* 1952, 2: 209.

37. Ibid.

38. Denham, "Surveys of the Birmingham chronic sick hospitals," 286–87; C. R. Lowe and T. McKeown, "Investigation of the need for physical treatment during prolonged absence from work," *British Journal of Preventive and Social Medicine* 7 (1953): 26–30.

39. On Ryle, see D. Porter, *Health citizenship: essays on social medicine and bio-medical politics* (Berkeley, 2012), 132–53.

40. T. McKeown, J. M. Mackintosh, and C. R. Lowe, "Influence of age on type of hospital to which patients are admitted," *Lancet* 277 (1961): 820.

41. F. N. Garratt, C. R. Lowe, and T. McKeown, "Investigation of the medical and social needs of patients in mental hospitals," *British Journal of Preventive & Social Medicine* 11 (1957): 165–73; idem, "Institutional care of the mentally ill," *Lancet* 271 (1958): 682–84.

42. T. McKeown, "The concept of a balanced hospital community," *Lancet* 271 (1958): 701–4.

43. Ibid., 702–3.

44. E. Wilkes, A. G. Crowther, and C. W. Greaves, "A different kind of day hospital for patients with pre-terminal cancer and chronic disease," *BMJ* 1978, 2: 1053–56.

45. R. M Green, "Beyond 'the Role of Medicine': McKeown as medical philosopher," *Milbank Q* 55 (1977): 392.

46. J. A. Ryle, *Changing disciplines* (London, 1948), 11–12, cited in M. Terris, "The Society for Epidemiologic Research (SER) and the future of epidemiology," *American Journal of Epidemiology* 136 (1992): 909.

47. J. A. Ryle and W. T. Russell, "The natural history of coronary disease: a clinical and epidemiological study," *British Heart Journal* 11 (1949): 370–89.

48. W.P.D. Logan, "Mortality in England and Wales from 1848 to 1947," *Population Studies* 4 (1950): 132–78. See especially 132–33.

49. W.P.D. Logan and E. M. Brooke, *The Survey of Sickness, 1943 to 1952* (London, 1957).

50. T. McKeown, "American medical services," *British Journal of Social Medicine* 2 (1948): 77–105.

51. S. Murphy, "The early days of the MRC Social Medicine Research Unit," *Soc Hist Med* 12 (1999): 398.

52. K. L. White, "Jerry Morris and health services research in the USA," *Int J Epid* 31 (2002): 690–92.

53. S. J. Kunitz, "Explanations and ideologies of mortality patterns," *Population and Development Review* 13 (1987): 379–408; J. A. Mendelsohn, "From eradication to equilibrium: how epidemics became complex after World War I," in *Greater than the parts: holism in biomedicine, 1920–1950*, ed. C. Lawrence and G. Weisz (New York, 1998), 303–30.

54. Murphy, "Early days of the MRC Social Medicine Research Unit"; J. Pemberton, "Origins and early history of the Society for Social Medicine in the UK and Ireland," *Journal of Epidemiology and Community Health* 56 (2002): 342–46.

55. D. Porter, "From social structure to social behaviour in Britain after the Second World War," *Contemporary British History* 16 (2002): 70.

56. V. Berridge, "Medicine and the public: the 1962 Report of the Royal College of Physicians and the new public health," *Bull Hist Med* 81 (2007): 288.

57. V. Berridge and K. Loughlin, "Smoking and the new health education in Britain 1950s–1970s," *AJPH* 95 (2005): 956–64.

58. E.g., D. Porter, "Calculating health and social change: an essay on Jerry Morris and late-modernist epidemiology," *Int J Epid* 36 (2007): 1180–84.

59. Quoted in Berridge, "Medicine and the public," 296; emphasis mine. Also see idem, "Jerry Morris," *Int J Epid* 30 (2001): 1141–45.

60. J. N. Morris, "Uses of epidemiology," *BMJ* 1955, 2: 395–401.

61. J. N. Morris, *Uses of epidemiology* (Edinburgh, 1957), 98.

62. J. N. Morris, "Some current trends in public health [and discussion]," *Proceedings of the Royal Society of London. Series B, Biological Sciences* 159 (1963): 73.

63. Ibid., 74.

64. J. N. Morris, *Uses of epidemiology*, 2nd ed. (Edinburgh, 1964), 278.

65. E.g., *Annual report of the Medical Officer of Health for the City and Port of Gloucester, for the year 1965* (Gloucester, 1966). Extracts were kindly provided by Martin Gorsky.

66. J. N. Morris, "The prevention of disease in middle age," *Public Health* 77 (1963): 237.

67. The classic critique is J. Lewis, *What price community medicine? The philosophy, practice and politics of public health since 1919* (Brighton, 1986).

68. J. N. Morris, "Tomorrow's community physician," *Lancet* 294 (1969): 811–16.

69. G. D. Smith, "The end of the beginning for chronic disease epidemiology," *Int J Epid* 39 (2010): 1–3.

70. Written communication provided by Martin Gorsky.

71. "Prevention of chronic disease," *BMJ* 1961, 2: 1137.

72. C. A. Boucher, "The young chronic sick," *Public Health* 81 (1967): 308–12.

73. "Annual Conference of Representatives of Local Medical Committees," *BMJ* 1968, 2: 202. Also see "Prescription charges," *BMJ* 1968, 1: 24–25, and "Annual Report of the Council 1967–1968," *Supplement to BMJ* 1968, 1: 83.

74. Ministry of Health, "Public expenditure: post-devaluation measures, prescription charges," Cabinet Paper c(68)14; 11 January 1968; Catalogue Ref CAB/129/135, p. 2, www .nationalarchives.gov.uk/documentsonline/details result.asp?queryType=1&resultcount =1&Edoc_Id=7980377.

75. "Prescription charges," 131.

76. Department of Health and Social Security, *Annual report of the Department of Health and Social Security for the year 1969* (London, 1969).

77. "'Inequitable and anomalous' prescription charges should be reviewed," *BMJ* 292 (1986): 709; S. Chaplin, "The NHS prescription charge: soon just for England," *Prescriber* 20 (2009): 25–32.

78. See the discussion within the CCI of American disability pensions in chapter 5 and the French case discussed in chapter 10.

79. N. S. Galbraith, "A national public health service," *J R Soc Med* 74 (1981): 17. On the failure of health promoters to professionalize, see P. Duncan, "Failing to professionalise, struggling to specialise: the rise and fall of health promotion as a putative specialism in England, 1980–2000," *Medical History* 57 (2013): 377–96.

80. Department of Health and Social Security, *Inequalities in health: report of a working group chaired by Sir Douglas Black* (London, 1980).

81. V. Berridge, "The Black report and the health divide," *Contemporary British History* 16 (2002): 131–72. Among the many articles resulting from the Whitehall studies, see G. Rose and M. G. Marmot. "Social class and coronary heart disease," *British Heart Journal* 45 (1981): 13–19.

82. J. N. Morris, "Epidemiology and prevention," *Milbank Q* 60 (1982): 1–16.

83. D. Acheson, *Public health in England: the report of the Committee of Inquiry into the Future Development of the Public Health Function* (London, 1988).

84. "Memorandum by the UK Public Health Association (UKPHA) (WP 105) to the Select Committee on Health, 2 June 2005," www.publications.parliament.uk/pa/cm200405/cmselect/cmhealth/358/358we113.htm.

85. "Annual Meeting BMA, Swansea, Debate: the routine medical checkup," *BMJ* 1965, 2: 289–90.

86. R. I. Bayliss, "The medical check-up," *BMJ* 283 (1981): 631–34. Critical letters are on pp. 792, 985–86, and 1124.

87. D. Armstrong, "Screening: mapping medicine's temporal spaces," *Sociology of Health & Illness* 34 (2012): 184, figure 4.

88. C. L. Leeming-Latham, "The seed and the soil: tuberculosis prevention and treatment in England, 1940–1974" (PhD diss., Birkbeck College, University of London, 2012), ch. 2.

89. J.M.G. Wilson, "Multiple screening," *Lancet* 282 (1963): 51–54.

90. R. J. Donaldson, "Multiple screening clinics," *Public Health* 81 (1967): 21.

91. "Surveillance and early diagnosis in general practice," *BMJ* 1965, 2: 472–74.

92. R. W. Elliott, "The prevention of illness in middle age," *Public Health* 79 (1965): 324.

93. A. L. Cochrane, "A medical scientist's view of screening," *Public Health* 81 (1967): 207–13.

94. J.M.G. Wilson, "Screening in the early detection of disease," in *Portfolio for health: the role and programme of the DHSS in health services research*, ed. G. McLachlan (London, 1971), 57–58.

95. W. W. Holland et al., "A controlled trial of multiphasic screening in middle-age: results of the South-East London Screening Study," *Int J Epid* 6 (1977): 357–63.

96. Wilson, "Screening in the early detection of disease," 54; *Breast cancer screening: report to the Health Ministers of England, Wales, Scotland and Northern Ireland by a working group chaired by Professor Sir Patrick Forrest 1986*, www.cancerscreening.nhs.uk/breast screen/publications/forrest-report.html; J. Austoker, "Cancer prevention in primary care: screening for cervical cancer," *BMJ* 309 (1994.): 241–48.

97. R.C.M. Pearson and A. Yarrow, "IV. Educating the public," *Public Health* 85 (1971): 67–75.

98. Galbraith, "A national public health service," 20.

99. *Public Health in England: the report of the Committee of Inquiry into the Future Development of the Public Health Function* (London, 1988), 15.

100. E.g., Unit for the Study of Health Policy, *Rethinking community medicine: report from a study group* (London, 1979).

101. J. Lewis, "The changing fortunes of community medicine," *Public Health* 100 (1986): 3–10.

102. J. Anderson, *War, disability and rehabilitation in Britain: "soul of a nation"* (Manchester, 2011), 180–82.

103. S. R. Tunbridge, *Report of a Subcommittee of the Standing Medical Advisory Committee, Central Health Services Council* (London, 1972); S. Mattingly, *Rehabilitation today* (London, 1977); M. Blaxter, "The future of rehabilitation services in Great Britain," *Disability and Rehabilitation* 2 (1980): 199–209; D. Gloag, "Rehabilitation of the elderly: 1—settings and services," *BMJ* 290 (1985), 455–57; M. A. Chamberlain, "The metamorphosis of physical medicine?" *J R Soc Med* 85 (1992): 131–35; R. Grahame, "The decline of rehabilitation services and its impact on disability benefits," *J R Soc Med* 95 (2002): 114–17.

104. E.g., Great Britain, Central Office of Information Reference Division, *Rehabilitation and care of the disabled in Britain* (London, 1969); J. Marshall and A. Haines, "Survey of the teaching of disability and rehabilitation to medical undergraduates in the UK," *Medical Education* 24 (1990): 528–30.

105. E.g., H. A. Rusk and M. M. Dacso, "The dynamic approach to the care of the chronic," *Hospitals* 29 (1955): 63–65; American Public Health Association, Program Area Committee on Chronic Disease and Rehabilitation, *Chronic disease and rehabilitation: a program guide for state and local health agencies* (New York, 1960).

106. A. E. Bennett, J. Garrad, and T. Halil, "Chronic disease and disability in the community: a prevalence study," *BMJ* 1970, 3: 762–64; J. Garrad and A. E. Bennett, "A validated interview schedule for use in population surveys of chronic disease and disability," *British Journal of Preventive & Social Medicine* 25 (1971): 97–104.

107. A. I. Harris, E. Cox, and C.R.W. Smith, *Handicapped and impaired in Great Britain* (London, 1971).

108. "Enabling the disabled," *Guardian*, May 2, 1973.

109. E. Badley, M. Roger, P. Thompson, and P. H. Wood. "The prevalence and severity

of major disabling conditions—a reappraisal of the Government Social Survey on the Handicapped and Impaired in Great Britain," *Int J Epid* 7 (1978): 151.

110. P. H. Wood, "Size of the problem and causes of chronic sickness in the young," *J R Soc Med* 71 (1978): 441.

111. E.g., M. Oliver and G. Zarb, "The politics of disability: a new approach," *Disability, Handicap & Society* 4 (1989): 221–39.

CHAPTER TEN: *Maladies chroniques* in France

1. H. Meding, *Paris médical: vade-mecum des médecins étrangers* (Paris, 1853), part 2, 20, 41–44, 142–46, 169; J. P. Domin, *Une histoire économique de l'hôpital, XIXe–XXe siècles: une analyse rétrospective du développement hospitalier*, 2 vols. (Paris, 2008), 1:142. (All references to this source are to volume 1.)

2. P. Guillaume, *Les hospices de Bordeaux au XIXe siècle, 1796–1855* (Bordeaux, 2000).

3. P. Guillaume, "Incurables et vieillards dans les hospices Bordelais," in *Lieux d'hospitalité: hospices, hôpital, hostellerie*, ed. Alain Montandon (Clermont-Ferrand, 2001), 133–34.

4. E.g., H. Couturier, *Projet d'un hôpital de convalescence: note adressée à MM. les membres du conseil d'administration des hospices civils de Lyon* (Lyon, 1866). On the treatment of incurable patients, see J. Szabo, *Incurable and intolerable: chronic disease and slow death in nineteenth-century France* (Rutgers, NJ, 2009).

5. France, Ministère de l'intérieur, *Rapport au roi sur les hôpitaux, les hospices et les services de bienfaisance* (Paris, 1837).

6. O. Faure, *Genèse de l'hôpital moderne: Les Hospices Civils de Lyon de 1802 à 1845* (Lyon, 1982).

7. Domin, *Une histoire économique*, 146–47.

8. Ibid., 142.

9. Szabo, *Incurable and intolerable*, 202.

10. C. Bec, "De la charité a l'assistance: le parcours Républicain (1880–1914)," in Musée de l'APHP, *Depuis cent ans, la société, l'hôpital et les pauvres* (Paris, 1996), 41.

11. Szabo, *Incurable and intolerable*, 208.

12. Ibid., 214.

13. Domin, *Une histoire économique*, 175.

14. Szabo, *Incurable and intolerable*, 210–15; G. Cros-Mayrevieille, *Traité de l'assistance hospitalière*, 3 vols. (Paris, 1912), 3:342–43. On the laicization of hospitals, see C. Chevandier, *L'hôpital dans la France du XXe siècle* (Paris, 2009), 63–101.

15. Cros-Mayrevieille, *Traité de l'assistance*, 343, 366–68.

16. Domin, *Une histoire économique*, 204.

17. T. B. Smith, "The social transformation of hospitals and the rise of medical insurance in France, 1914–1943," *Historical Journal* 41 (1998): 1055–87.

18. O. Faure and D. Dessertine, *Les cliniques privées: deux siècles de succès* (Rennes, 2012).

19. P. V. Dutton, *Origins of the French welfare state: the struggle for social reform in France, 1914–1947* (Cambridge, 2002).

20. Domin, *Une histoire économique*, 190.

21. *Hospices et hôpitaux: règlement intérieur: circulaire du 31 mars 1926* (Nancy, 1927).

22. Olivier Faure, "L'hôpital et les incurables au XIXe siècle: l'exemple de Lyon," *Handicaps et inadaptations* 50 (1990): 76–77; P. Guillaume, *Du désespoir au salut: les tuberculeux aux 19e et 20e siècles* (Paris, 1986); P. Pinell, *Naissance d'un fléau: histoire de la lutte contre le cancer en France, 1890–1940* (Paris, 1992).

23. Domin, *Une histoire économique*, 220.

24. E. Feller, *Histoire de la vieillesse en France, 1900–1960: du vieillard au retraité* (Paris, 2005), 120–34; P. Theil, *Le corps médical devant la médecine sociale* (Paris, 1943), 127.

25. *JAMA* 92 (1929): 160–61.

26. J. Couteaux, "Un hôpital de chroniques: Hôpital Raymond-Poincaré à Garches," *Revue Hospitalière de France* 3 (1938): 104–12; S. Riché and S. Riquier, *Des hôpitaux à Paris: état des fonds des archives de L'AP-HP, XIIe–XXe siècle* (Paris, 2000), 671–77; Lacaze, "Aménagement d'un centre de cure et de réadaptation pour les enfants de polio," *Commission de Surveillance des Hôpitaux de Paris*, Oct. 16, 1947, 61–63, and Nuemeyer, "Création d'un école de rééducation des malades osseux," ibid., April 8, 1948, 375–77, both in AAP-HP, 1L 82.

27. M. Finance, "Les services de maladies chroniques," report submitted to the CSH meeting of June 27, 1951, in ANF, AG 1347. (I am grateful to Marie-Odile Frattini for providing me with a copy of this document and those cited in the following three notes.)

28. R. F. Bridgman, "L'hospitalisation des vieillards malades," ANF, AG 1347.

29. All but the final report are in CSH, "Notes remise en séance de déc. 13, 1950," ANF, JM/20/50.

30. "Rapport de synthèse sur les malades chroniques dans le hôpitaux" (n.d.) ANF, AG 1347.

31. Administration Générale de l'Assistance Publique a Paris, *L'Hôpital de Créteil* (Paris, 1954); Riché and Riquier, *Des hôpitaux à Paris*, 247.

32. M. Perretti, "Fixation des prix de journée à appliquer dans les hôpitaux en 1962, Conseil D'Administration," AP, Dec. 12, 1961, 443; Dr. Bousser, "Programme des grand travaux d'équipement inscrits au 3e plan," ibid., Jan. 18, 1962, 535; "Communication de M. de Directeur General" and "Programme des Travaux," ibid., April 26, 1962, 752–55, 1108–16; "Réalisations d'un programme de construction de 2000 lits," May 15, 1962, 1162–68. All in AAP-HP, 1L99 and 1L100. On these hospitals, see Riché and Riquier, *Des hôpitaux à Paris*, 414–15, 464–69, 681–83.

33. A. Sentilhes-Monkam, "Rétrospective de l'hospitalisation à domicile: l'histoire d'un paradoxe," *Revue française des affaires sociales* 3 (2005): 163–64.

34. A. Bouchet, R. Mornex, and D. Gimenez, *Les hospices civils de Lyon: histoire de leurs hôpitaux* (Lyon, 2003), 7–8.

35. M. Grivaux, "De quelques problèmes hospitalières (V)," *Semaine des hôpitaux de Paris* 39 (1963): 1023; E. Boltanski, "A propos de la transformation des sanatoriums en services de chroniques," *Bull AM* 149 (1965): 604–5.

36. M. Mayen, "La population des hospices et des maisons de retraite," *Population (French Edition)* 27 (1972): 70.

37. A-M. Guillemard, *Le déclin du social: formation et crise des politiques de la vieillesse* (Paris, 1986), 159, 395; Mayen, "La population des hospices"; R. Grégoire, *Pour une politique de la santé: rapports présentés à Robert Boulin, Ministère de la santé publique et de la sécurité sociale, vol. 3: l'hôpital* (Paris, 1971), 90.

38. A. Chauvenet, *Médecines au choix, médecine de classes* (Paris, 1978), 50.

39. P. Aboulker, L. Chertok, and M. Sapir, *Psychosomatique et chronicité: 2e Congrès International de Médecine Psychosomatique de Langue Française, 5–8 juillet 1963* (Paris, 1964).

40. Grivaux, "De quelques problèmes hospitaliers (V)," 1027–28.

41. G. Weisz "A moment of synthesis: holism in inter-war France," in *Greater than the parts: holism in biomedicine 1920–1950*, ed. C. Lawrence and G. Weisz (New York, 1998), 68–93.

42. Note in *JAMA* 146 (1951b): 1060.

43. P. Delore, *L'hôpital humanisé* (Paris, 1959), 113.

44. P. Delore, *De la médecine clinique à la médecine sociale* (Paris, 1961), 107.

45. Ibid., 108.

46. Ibid., 117.

47. Ibid., 104.

48. D. Benamouzig, *La santé au miroir de l'économie* (Paris, 2005), 34–36.

49. H. Péquignot, G. Rösch, and Y. Salomon, "Le vieillissement dans un hospice de vieillards," *Presse Médicale* 65 (1957): 1574–75.

50. H. Péquignot and M. Gatard, *Hôpital et humanisation* (Paris, 1976), 46.

51. H. Péquignot and G. Rösch, *Notre vieillissement* (Paris, 1960), 99–102.

52. H. Péquignot, "Le problème des maladies chroniques," *Bull AM* 148 (1964): 733–36; idem, "Les maladies chroniques et l'organisation des soins médicaux," *Bull AM* 149 (1965): 190–93.

53. Péquignot, "Les maladies chroniques," 194–95.

54. Ibid., 195.

55. Much of what follows in the next two paragraphs is based on Guillemard, *Le déclin du social*. See also idem, *La vieillesse et l'état* (Paris, 1980).

56. S. De Beauvoir, *La vieillesse*, 2 vols. (Paris, 1970).

57. Péquignot and Rösch, *Notre vieillissement*; H. Péquignot, *Vieillir et être vieux* (Paris, 1981); idem, *Vieillesses de demain: vieillir et être vieux* (Paris, 1986).

58. C. Coatanea, *La transformation des hospices: bilan de la Loi du 30 juin 1975* (Rennes, 1987), 15–18.

59. P. Laroque and Haut Comité Consultatif de la Population et de la Famille, *Politique de la vieillesse: rapport de la commission d'étude des problèmes de la vieillesse* (Paris, 1962), 232–43. The same imbalance can be seen in the special issue of the influential social-Catholic journal *L'Esprit* 31 (1963), titled "Vieillesse et vieillissement."

60. *Pour une politique de la santé, 6, Aspects médicaux du vieillissement* (Paris, 1970), cited in N. Henckes and M. Bungener, "L'émergence du secteur médico-social en France, 1945–1990: entre planification, expertise médico-économique et la transformation de l'organisation des soins," unpublished paper, 17. For a similar process in disability policy, see ibid., 18. On the relationship between social and medical approaches, see N. Benoit-Lapierre, "Guérir de vieillesse," *Communications* 37 (1983): 149–65.

61. D. Sicart, *Série Statistiques—157—Les médecins au 1er janvier 2011, document de travail Ministère du Travail, de l'Emploi et de la Santé*, www.drees.sante.gouv.fr/IMG/pdf/seriestat157-2.pdf.

62. Coatanea, *La transformation des hospices*, 27–28; A. Catherin-Quivet, "Évolution de la population âgée en institution et politiques mises en œuvre (1962–2004)," *Annales de démographie historique* 110 (2005): 186.

63. R. Lenoir, *Les exclus: un Français sur dix* (Paris, 1974). On handicap and exclusion, see S. Ebersold, *L'invention du handicap: la normalisation de l'infirme* (Paris, 1992).

64. The literature on exclusion as a socioeconomic problem is immense. See, for instance, C. Martin and R. A. B. Leaper. "French review article: the debate in France over 'social exclusion,'" *Social Policy & Administration* 30 (1996): 382–92. A critical text is R. Castel, *Les métamorphoses de la question sociale: une chronique du salariat* (Paris, 1995).

65. "La réforme hospitalière voit le jour," *Technique Hospitalière* 307 (1971) : 57.

66. This is clearly articulated in Grégoire, *Pour une politique de la santé*, 41.

67. É. Molinié, Conseil économique et social, *L'hôpital public en France: bilan et perspectives* (Paris, 2005), 13–14.

68. Echelon National du Service Médical Groupe d'Animation et d'Impulsion national, *Les services de soins de suite ou de réadaptation: résultats nationaux* (Paris, 1998).

69. Coatanea, *La transformation des hospices*, 30.

70. Ibid., 39, 66.

71. Ibid., 25–26, 41, 50–57; Catherin-Quivet, "Évolution de la population âgée," 186; N. Benoit-Lapierre, R. Cevasco, and M. Zafiropoulos, *Vieillesse des pauvres: les chemins de l'hospice* (Paris, 1980); P. Hofman, Conseil économique et social, *La réforme hospitalière: bilan et perspectives: avis* (Paris, 1983), 32–45.

72. J. M. Clément, *Histoire des réformes hospitalières sous la Ve République* (Bordeaux, 2010).

73. Hofman, *La réforme hospitalière*, 57.

74. F. Tugores, "La clientèle des établissements d'hébergement pour personnes âgées, situation au 31 décembre 2003," *DREES: Etudes et résultats* 485 (2006): 1–7.

75. E.g., T. Frinault, "La dépendance ou la consécration française d'une approche ségrégative du handicap," *Politix* 72 (2005): 11–31.

76. M. Danzig, *Handicap et maladie évolutive: rapport d'un groupe d'études présidé par Mme. le Docteur M. Danzig, Mai 1976–Decembre 1977* (Paris, 1978); M. Winance, "Handicap et normalisation: analyse des transformations du rapport à la norme dans les institutions et les interactions," *Politix* 17 (2004): 201–27.

77. E.g., B. Ennuyer, "1962–2007: Regards sur les politiques du 'maintien a domicile,'" *Gérontologie et société* 4 (2007): 153–67.

78. Catherin-Quivet, "Évolution de la population âgée," 195.

79. Sentilhes-Monkam, "Rétrospective de l'hospitalisation à domicile," 162; Hofman, *La réforme hospitalière*, 50–52.

80. A. Afrite, M. Chaleix, L. Com-Ruelle, and H. Valdelievre, "L'hospitalisation à domicile, une prise en charge qui s'adresse à tous les patients: exploitation des données du PMSI HAD 2006," *Questions d'économie de la santé* 140 (2009): 1–8.

81. L. Murard and P. Zylberman, *L'hygiène dans la République: la santé publique en*

France, ou l'utopie contrariée, 1870–1918 (Paris, 1996); idem, "Mi- ignoré, mi- méprisé: le Ministère de la Santé Publique, 1920–1945," *Les tribunes de la santé* 1 (2003): 19–33.

82. N. Henckes, "Reforming psychiatric institutions in the mid-twentieth century: a framework for analysis," *History of Psychiatry* 22 (2011): 170.

83. Commissariat General du Plan, *Rapport général de la Commission de la Santé et de l'Assurance Maladie*, March 1976, 36.

84. I. Durand-Zaleski and M. D. Campion, *Les politiques de prévention* (Paris, 2003).

85. L. Berlivet, "Between expertise and biomedicine: public health research in France after the Second World War," *Medical History* 52 (2008): 471–92; M. Menoret, "The genesis of the notion of stages in oncology: the French Permanent Cancer Survey (1943–1952)," *Soc Hist Med* 15 (2002): 291–302.

86. Henckes and Bungener, "L'émergence du secteur médico-social."

87. Ibid., 3.

88. Ibid., 12.

89. Berlivet, "Between expertise and biomedicine"; J. F. Picard, "Aux origines de l'Inserm: André Chevallier et l'Institut National d'Hygiène," *Sciences sociales et santé* 21 (2003): 5–26.

90. Hofman, *La réforme hospitalière*, 82.

91. I. Aujoulat, A-L. Le Faou, B. Sandrin-Berthon, F. Martin, and A. Deccache, "Implementing health promotion in health care settings: conceptual coherence and policy support," *Patient Education and Counseling* 45 (2001): 245–54.

92. J. P. Deschamps, "Assurance-maladie et prévention en France: a propos des examens systématiques de sante," *Revue française des affaires sociales* 49 (1995), 135–41; J. J. Moulin, V. Dauphinot, C. Dupré, C. Sass, E. Labbe, L. Gerbaud, and R. Guéguen, "Inégalités de santé et comportements: comparaison d'une population de 704128 personnes en situation de précarité à une population de 516607 personnes non précaires, France, 1995–2002," *Bulletin épidémiologique hebdomadaire* 43 (2005): 1–3.

93. Commissariat General du Plan, *Rapport général 1976*, 36.

94. V. Vignaux, "L'éducation sanitaire par le cinéma dans l'entre-deux-guerres en France," *Sociétés & Représentations* 2 (2009): 67–85.

95. Commissariat General du Plan, *Rapport général 1976*, 37; L. Berlivet, "Information is not good enough: the transformation of health education in France in the late 1970s," *Journal of Epidemiology and Community Health* 62 (2008): 7–10; L. Berlivet, "Une biopolitique de l'éducation pour la santé," in *Le gouvernement des corps*, ed. D. Memmi and D. Fassin (Paris, 2004), 37–75.

96. J. L. Dhondt, J. P. Farriaux, M. L. Briard, R. Boschetti, and J. Frézal, "Neonatal screening in France," *Screening* 2 (1993): 77–85.

97. P. Schaffer, H. Sancho-Garnier, M. Fender, P. Dellenbach, J. P. Carbillet, E. Monnet, G. P. Gauthier, and A. Garnier, "Cervical cancer screening in France," *European Journal of Cancer* 36 (2000): 2215–20; Société Française de Médecine Préventive et Sociale, *Rapport sur la nécessité d'extension du dépistage des cancers en France: 36ᵉ Journée Scientifique de la Société Française de Médecine Préventive et Sociale* (Paris, 1992).

98. M. Tubiana, "Préface," in *Dépistage des cancers: de la médecine à la santé publique*, ed. H. Sancho-Garnier et al. (Paris, 1997), 1.

99. Agence Nationale d'Accréditation et d'Evaluation en Sante, *Evaluation du Programme National de Dépistage Systématique du Cancer du Sein, Mars 1997* (Paris, 1997); H. Sancho-Garnier, "Enquête sur les difficultés de mise en place du dépistage des cancers du sein," in *Séminaire résultats des campagnes de dépistage du cancer du sein, 27–28 Octobre 1997: compte-rendu*, ed. R. Ancelle-Park (Marseille, 1997), 39–41; R. Ancelle-Park, J. Nicolau, and A. C. Paty, "Programme de Dépistage Organisé du Cancer du Sein: tendances des indicateurs précoces," *Bulletin épidémiologique hebdomadaire* 4 (2003): 14–16.

100. P. Morel, "PROFESSION-Dépistage des cancers: la fin du retard français? Ou en est-on de la généralisation promise?" *Concours Médical* 124 (2002): 2252–55.

101. C. F. Roques and J. F. Mathé, "Physical medicine and rehabilitation in France," *Disability & Rehabilitation* 19 (1997): 366–70.

102. F. Bloch-Lainé, *Etude du problème général de l'inadaptation des personnes handicapées: rapport présenté au Premier Ministre* (Paris, 1969).

103. Roques and Mathé, "Physical medicine and rehabilitation," 366–70.

104. J. P. Devailly and L. Josse, "Accès aux soins de réadaptation et handicap," *Gestions hospitalières* 492 (2010): 28–33. M-O. Frattini, "Dynamique de constitution d'une spécialité médicale fragile: la médecine de rééducation et réadaptation fonctionnelles en France entre médecine et politique," Ecoles des Hautes Etudes en Science Sociales, Mémoire de Master 2, 2008, suggests that emphasis on functionality has been prevalent since the 1950s.

Epilogue

1. K. G. Alberti, "Interhealth: the WHO integrated programme for community health in non-communicable diseases," *J R Coll Phys* 27(1993): 65–69.

2. X. Berrios et al., "Distribution and prevalence of major risk factors of noncommunicable diseases in selected countries: the WHO Inter-Health Programme," *Bull WHO* 75 (1997): 99–108.

3. E. Nolte and M. McKee, "Integration and chronic care: a review," in *Caring for people with chronic conditions: a health system perspective*, ed. E. Nolte and M. McKee (Maidenhead, 2008), 66.

4. "International Symposium on Health Promotion and Chronic Illness, Bad Honnef, Federal Republic of Germany, 21–25 June 1987," *Health Promotion International* 2 (1987): 387.

5. D. J. Hunter and G. Fairfield, "Disease management," *BMJ* 315 (1997): 50–53.

6. D. M. Bott, M. C. Kapp, L. B. Johnson, and L. M. Magno, "Disease management for chronically ill beneficiaries in traditional Medicare," *Health Affairs* 28 (2009): 86–98.

7. Ibid.; M. B. Rosenman et al., "The Indiana Chronic Disease Management Program," *Milbank Q* 84 (2006): 135–63.

8. E. H Wagner, C. Davis, J. Schaefer, M. Von Korff, and B. Austin, "A survey of leading chronic disease management programs: are they consistent with the literature?" *Managed Care Quarterly* 7 (1999): 56–66; T. Bodenheimer, E. H. Wagner, and K. Grumbach, "Improving primary care for patients with chronic illness," *JAMA* 288 (2002): 1775–79, 1909–14.

9. E. H. Wagner, "Chronic disease management: what will it take to improve care for chronic illness?" *Effective Clinical Practice* 1 (1998): 2–4.

10. E.g., A. K. Parekh and M. B. Barton, "The challenge of multiple comorbidity for the US health care system," *JAMA* 303 (2010): 1303–4.

11. Bodenheimer, Wagner, and Grumbach, "Improving primary care," 1775–79.

12. C. Ham, "The ten characteristics of the high-performing chronic care system," *Health Economics, Policy and Law* 5 (2010): 80.

13. Subcommittee on Health, *Eliminating barriers to chronic care management in Medicare: hearing before the Subcommittee on Health of the Committee on Ways and Means, U.S. House of Representatives, One Hundred Eighth Congress, first session, February 25, 2003* (Washington, DC, 2003); Rosenman et al., "The Indiana Chronic Disease Management Program."

14. A. Bitton, C. Martin, and B. E. Landon, "A nationwide survey of patient centered medical home demonstration projects," *Journal of General Internal Medicine* 25 (2010): 584–92; R. A. Berenson, *Challenging the status quo in chronic disease care: seven case studies* (Oakland, CA, 2006).

15. J. E. Epping-Jordan, S. D. Pruitt, R. Bengoa, and E. H. Wagner, "Improving the quality of health care for chronic conditions," *Quality and Safety in Health Care* 13 (2004): 299–305.

16. WHO, *Preparing a health care workforce for the 21st century: the challenge of chronic conditions* (Geneva, 2005).

17. E.g., J. Lundsgaard, "Consumer direction and choice in long-term care for older persons, including payments for informal care: how can it help improve care outcomes, employment and fiscal sustainability?" *OECD health working papers* (Paris, 2005); T. Frits, M. Jérôme, L. N. Ana, and C. Francesca, *OECD health policy studies help wanted? Providing and paying for long-term care* (Paris, 2011).

18. According to Robert Beaglehole, a major figure in the global health movement, this shift was the result of an effort to follow United Nations terminology. Personal communication, Feb. 24, 2013.

19. E.g., WHO, *Package of essential noncommunicable (PEN) disease interventions for primary health care in low-resource settings* (Geneva, 2010).

20. Members are the National Health and Medical Research Council of Australia, the Canadian Institutes of Health Research, the Medical Research Councils of the United Kingdom and South Africa, the National Heart, Lung, and Blood Institute of the National Institutes of Health in the United States, and the Chinese Academy of Medical Sciences. www.nhmrc.gov.au/grants/apply-funding/global-alliance-chronic-diseases. Accessed Feb. 14, 2013.

21. J. Marcus, "Silent epidemic," *New Republic*, April 29, 2011, www.tnr.com/article/health-care/87595/moscow-noncommunicable-diseasesafrica-cancer.

22. "UN gathering on non-communicable diseases considers ways to combat scourge," *UN News Centre*, September 20, 2011.

23. "Development of an updated action plan for the global strategy for the prevention and control of noncommunicable diseases covering the period 2013 to 2020," WHO discussion paper, July 26, 2012, www.who.int/nmh/events/2012/action_plan_20120726.pdf; "Draft action plan for the prevention and control of noncommunicable diseases 2013–

2020: Report by the Secretariat," January 11, 2013, EB132/7, http://apps.who.int/gb/ebwha/pdf_files/EB132/B132_7-en.pdf.

24. Frits, Jérôme, and Francesca, *OECD health policy studies*.

25. *BMJ* 320 (2000). For the British view, see especially D. J. Hunter, "Disease management: has it a future?" ibid., 530.

26. C. Ham, "Chronic care in the English National Health Service: progress and challenges," *Health Affairs* 28 (2009): 192; R. Rosen, "Developing chronic disease policy in England," in *Emerging approaches to chronic disease management in primary health care*, ed. J. Dorland and M. A. McColl (Montreal, 2007), 39–50.

27. Department of Health, Great Britain, *Supporting people with long term conditions: an NHS and social care model to support local innovation and integration* (London, 2005), www.dh.gov.uk/prod_consum_dh/groups/dh_digitalassets/@dh/@en/documents/digitalasset/dh_4122574.pdf.

28. D. de Silva and D. Fahey, "England," in *Managing chronic conditions: experience in eight countries*, ed. E. Nolte, C. Knai, and M. McKee (Copenhagen, 2008), 30–31, 44: Table 2a. NHS HCHS: Nursing, Midwifery & Health Visiting staff and support staff by type 2002–2012 in NHS Hospital and Community Health Service (HCHS) Workforce Statistics in England, Non-medical staff—002–2012, as at 30 September, Publication date: March 21, 2013. www.hscic.gov.uk/searchcatalogue?productid=11216&q=community+matrons&sort=Relevance&size=10&page=1#top.

29. Ham, "Chronic care in the English NHS," 193.

30. Rosen, "Developing chronic disease policy," 39–50.

31. De Silva and Fahey, "England," 33.

32. Ibid., 45; Ham, "Chronic care in the English NHS," 198.

33. Department of Health, Great Britain, *Long term conditions compendium of information, 3rd edition* (2012), www.dh.gov.uk/prod_consum_dh/groups/dh_digitalassets/@dh/@en/documents/digitalasset/dh_134486.pdf.

34. LOI n° 2004–806 du 9 août 2004 relative à la politique de santé publique (Journal Officiel (JO) du 11 août 2004] Annexe, www.legifrance.gouv.fr/jopdf/common/jo_pdf.jsp?numJO=0&dateJO=20040811&numTexte=4&pageDebut=14277&pageFin=14337.

35. Loi n°2004–810 du 13 août 2004 relative à l'assurance maladie, www.legifrance.gouv.fr/affichTexte.do?cidTexte=JORFTEXT000000625158.

36. France, Ministère de Solidarités de la Sante et de la Famille, Direction Général de la Santé, *Actes du séminaire préparatoire au plan visant à améliorer la qualité de vie des personnes atteintes de maladies chroniques, 8 décembre 2004*, Nancy, 2004, www.sante.gouv.fr/IMG/pdf/plan_actes2005–2.pdf.

37. Les Chroniques Associés, "Maladies chroniques et invalidité," Sept. 2008, www.chroniquesassocies.fr/plaquettes/chroniques/Asso_Invalid_vF.pdf.

38. P. L. Bras, G. Duhamel, and E. Grass, *Améliorer la prise en charge des malades chroniques: les enseignements des expériences étrangères de "disease management"; Rapport RM2006-136P, IGAS* (Paris, 2006).

39. Ministère de la Santé et des Solidarités. *Plan pour l'amélioration de la qualité de vie des personnes atteintes de maladies chroniques, 2007–2011*, April 2007.

40. Rapport annuel du comité de suivi 2011 (juin 2012), www.sante.gouv.fr/plan-pour-l-amelioration-de-la-qualite-de-vie-des-personnes-atteintes-de-maladies-chroniques-2007-2011.html.

41. http://chronisante.inist.fr.

42. For instance, the Plan Alzheimer 2008–12 has as one of it three axes improving the quality of life of patients and their caregivers. www.plan-alzheimer.gouv.fr/. There are similar measures in the more complex Plan Cancer 2009–13. www.plan-cancer.gouv.fr/le-plan-cancer/5-axes-30-mesures/axe-soins/mesure-20.html.

43. S. de Chambine and A. Morin, "Protocolisation et qualité du parcours de soins dans le dispositif des affections de longue durée," *Actualité et Dossier en Sante Publique* 72 (2010): 25–38; "Promouvoir les parcours de soins personnalisés pour les malades chroniques" www.has-sante.fr/portail/jcms/c_1247611/promouvoir-les-parcours-de-soins-personnalises-pour-les-malades-chroniques. Accessed March 2, 2013.

44. Cour des comptes: Rapport public annuel 2013—février 2013, ch. 3.

45. www.healthways.com/worldwide/worldwide.aspx; (2008–2013) Sophia, France, évaluation de la prise en charge du diabète par l'Inspection générale des affaires sociales (IGAS), http://chronisante.inist.fr/?2012-Sophia-France-evaluation-de. Both accessed Feb. 25, 2013. Ch. G. et S. L "Anti-Sophia, des médecins perdent leur sang-froid," *Quotidien du Médecin*, March 2, 2013, www.lequotidiendumedecin.fr/actualite/exercice/anti-sophia-des-medecins-perdent-leur-sang-froid?ku=8C7wa958-va5a-C7wD-Ez8y-6Cyw BADw6AAA.

46. M. Rijken et al., "Chronic disease management programmes: an adequate response to patients' needs?" *Health Expectations* (Early view, 2012), http://nvl002.nivel.nl/postprint/PPpp4172.pdf.

47. G. Weisz, A. Cambrosio, P. Keating, L. Knaapen, T. Schlich, and V. J. Tournay, "The emergence of clinical practice guidelines," *Milbank Q* 85 (2007): 691–727.

Index